T0254615

SpringerBriefs in Electrical and Computer Engineering

SpringerBriefs present concise summaries of cutting-edge research and practical applications across a wide spectrum of fields. Featuring compact volumes of 50 to 125 pages, the series covers a range of content from professional to academic. Typical topics might include: timely report of state-of-the art analytical techniques, a bridge between new research results, as published in journal articles, and a contextual literature review, a snapshot of a hot or emerging topic, an in-depth case study or clinical example and a presentation of core concepts that students must understand in order to make independent contributions.

More information about this series at http://www.springer.com/series/10059

Igor Bolvashenkov · Hans-Georg Herzog
Ilia Frenkel · Lev Khvatskin
Anatoly Lisnianski

Safety-Critical Electrical Drives

Topologies, Reliability, Performance

Igor Bolvashenkov
Institute of Energy Conversion Technology
Technical University of Munich (TUM)
Munich
Germany

Lev Khvatskin
Center for Reliability and Risk Management
SCE-Shamoon College of Engineering
Beersheba
Israel

Hans-Georg Herzog
Institute of Energy Conversion Technology
Technical University of Munich (TUM)
Munich
Germany

Anatoly Lisnianski
The System Reliability Department
Israel Electric Corporation Ltd.
Haifa
Israel

Ilia Frenkel
Center for Reliability and Risk Management
SCE-Shamoon College of Engineering
Ashdod
Israel

ISSN 2191-8112 ISSN 2191-8120 (electronic)
SpringerBriefs in Electrical and Computer Engineering
ISBN 978-3-319-89968-8 ISBN 978-3-319-89969-5 (eBook)
https://doi.org/10.1007/978-3-319-89969-5

Library of Congress Control Number: 2018938371

Printed on acid-free paper

This Springer imprint is published by the registered company Springer International Publishing AG part of Springer Nature
The registered company address is: Gewerbestrasse 11, 6330 Cham, Switzerland

Preface

The problem of vehicular electrification has recently become an extremely important engineering task. The most significant element in the solution of this problem is the development of highly efficient and fault-tolerant traction electric drives that will be the optimal choice for vehicles performing under specified operating conditions. Since vehicular traction electric drives are safety-critical systems, they are subject to the most stringent requirements with respect to reliability and fault tolerance.

The aim of this book is to provide a comprehensive presentation of multi-criteria operational efficiency analysis and multifactor reliability-oriented evaluation of the safety-critical traction electrical drive that would be most suitable for icebreaker ships, taking into account the real conditions of ice navigation.

To solve the above task, we applied new theoretical approaches, including combined random process methods, the *Lz*-transform technique for multistate systems, statistical data processing, and Markov reward models for the calculation of the reliability-associated cost of multistate systems (MSS).

The authors anticipate that this book will be of considerable interest to researchers, practical engineers, and industrial managers who are involved in the addressing of issues related to the design and operation of safety-critical traction electric drives. In addition, it will be a helpful textbook for undergraduate and graduate courses in several departments, including electrical engineering, industrial engineering, mechanical engineering, and applied mathematics. This book is self-contained and does not require the reader to use other books or papers.

There are four chapters in this book.

Chapter 1 introduces the electric propulsion systems of icebreaker Arctic ships as an object of study. It presents the methodology of multi-criteria comprehensive operational efficiency estimation and describes the basic statistics of Arctic icebreaker ships with competitive traction drives. This chapter also discusses the impact of operating conditions on the applicability and preference of the traction electrical drive compared with the mechanical gear drive.

Chapter 2 evaluates the highly important reliability features of icebreaker ships, which were discussed in the previous chapter, through the use of *Lz*-transform, the modern stochastic process method for the reliability assessment of an MSS. Basing ourselves on the stochastic model while considering the conditions of ice navigation in the Western sector of the European part of the Arctic during winter and summer periods, we determined the operational availability of icebreaker ships, their power performance, and their power performance deficiency.

Chapter 3 focuses on solving the problem of choosing the optimal type of main traction electric motor for the electric propulsion system in two types of Arctic ships, an icebreaker cargo ship and an icebreaker, on the basis of the technique described in Chap. 1. We investigated the competitive options of the traction electric motor, the induction motor, and the permanent magnet synchronous motor, and we conducted a comparative analysis of two options, basing ourselves on the statistical data on the Arctic operations of icebreakers.

Chapter 4 is concerned with reliability-associated cost evaluation as a component in the process of decision-making in the choice of the optimal type of traction electric motor, which is described in Chap. 3. The chapter introduces reliability-associated cost as an important part of the life cycle expenses for any repairable MSS, taking into account the aging process of components. It describes Markov reward models as a basic tool for the computation of reliability-associated cost.

The Appendix presents corresponding MATLAB® codes.

Munich, Germany — Igor Bolvashenkov
Munich, Germany — Hans-Georg Herzog
Ashdod, Israel — Ilia Frenkel
Beersheba, Israel — Lev Khvatskin
Haifa, Israel — Anatoly Lisnianski

Contents

About the Authors

Igor Bolvashenkov, Ph.D., is a Senior Researcher at the Institute of Energy Conversion Technology of Technical University of Munich (TUM), Munich, Germany. He obtained his M.Sc. (1981) and Ph.D. degrees (1989) in Electrical Engineering from Admiral Makarov State University of Maritime and Inland Shipping, Leningrad, USSR. From 1987 to 1993, he worked as an Associate Professor at the Murmansk State Technical University, Russia. Since 2004, he has worked at the Institute of Energy Conversion Technology at the Technical University of Munich (TUM), Munich, Germany.

He specializes in the development and simulation of electric propulsion safety-critical system for ships, cars, and aircraft and analysis of their reliability, survivability, and fault tolerance.

He has published more than 90 scientific articles, chapters, and patents.

Prof. Dr.-Ing. Hans-Georg Herzog works at the Institute of Energy Conversion Technology, Technical University of Munich (TUM), Munich, Germany. He holds a diploma and doctoral degree (with distinction) from the Technical University of Munich (TUM). After his time as a research associate, he joined Robert Bosch GmbH, Leinfelden-Echterdingen, Germany. Since 2002, he has been Head of the Institute of Energy Conversion Technology at TUM.

His main research interests are energy efficiency of hybrid electric and full electric vehicles, electric aircraft, reliability of drive trains and their components, energy and power management, and advanced design methods for electrical machines. He is a Senior Member of IEEE and Member of VDI as well as VDE.

Ilia Frenkel, Ph.D., is Chair of the Center for Reliability and Risk Management and Senior Lecturer of the Industrial Engineering and Management Department, Shamoon College of Engineering (SCE), Ashdod, Israel.

He obtained his M.Sc. degree in Applied Mathematics from Voronezh State University, Russia, and Ph.D. degree in Operational Research and Computer Science from the Institute of Economy, Ukrainian Academy of Science, Kiev, Ukraine. He has more than 40 years of academic experience and teaching experience at universities and institutions in Russia and Israel. From 1988 till 1991, he worked as Department Chair and Associate Professor at the Applied Mathematics and Computers Department, Volgograd Civil Engineering Institute, Russia. From 2005, he served as Chair of the Center for Reliability and Risk Management and Senior Lecturer at the Industrial Engineering and Management Department, Shamoon College of Engineering (SCE), Ashdod, Israel.

He specializes in applied statistics and reliability with application to preventive maintenance. He is an editor and a member of the editorial board of scientific and professional journals.

He published one book and more than 50 scientific articles and chapters. He has edited 5 Books and 12 Special Journal Issues.

Lev Khvatskin, Ph.D., is a Researcher of the Center for Reliability and Risk Management and Senior Lecturer of the Industrial Engineering and Management Department, Shamoon College of Engineering (SCE), Beersheba, Israel. He obtained his M.Sc. degree in Mechanical Engineering and his Ph.D. degree in Reliability Engineering and Inventory Control from the University of Railway Transport, Moscow, Russia.

He specializes in statistical and probabilistic methods in reliability of power and refrigeration equipment, scheduled and unscheduled repair and service, refrigeration, heating and air-conditioning systems.

He published more than 30 scientific articles and chapters in the fields of reliability, applied statistics and production and operation management.

Anatoly Lisnianski, Ph.D., is a Senior Engineering Expert at the System Reliability Department of the Israel Electric Corporation Ltd., Israel, and Scientific Supervisor of the Center for Reliability and Risk Management, Shamoon College of Engineering (SCE), Beersheba, Israel.

He received his M.Sc. degree in Electrical Engineering from the Saint Pertersburg State University of Information Technologies, Mechanics and Optics, and his Ph.D. degree in Reliability in 1984 from the Federal Scientific and Production Center "Aurora" in Saint Petersburg, Russia. From 1975 through 1989, he was a researcher at this center. Since 1991, he has worked as an Engineering Expert at the Reliability Department of the Israel Electric Corporation. He specializes in reliability assessment and optimization for complex technical systems and is the author and co-author of more than 120 journal papers, two books, a number of chapters, and has patents in the field of reliability and applied statistics. He is a Senior Member of IEEE and a Member of Israel Statistical Association.

Chapter 1
Choosing the Optimal Type of Safety-Critical Traction Drives for Arctic Ships Based on Estimated Operational Efficiency and Real Ice Navigation Conditions

Abstract This chapter presents the methodology and results of a comparative analysis of the operational efficiency of icebreaker ships with electrical and mechanical transmission of power to the propeller for the Arctic region. This study has been carried out based on statistical operational data and stochastic models. Our conclusion is founded on the suitability of one of the traction drives we compared and which are employed in ships designed for Arctic operating conditions.

Keywords Icebreaker ships · Diesel-electric drive · Diesel-geared drive
Arctic operation · Reliability-associated cost · Operational efficiency
Multi-power source traction drive

1.1 Introduction

The modern development of new technologies, equipment and materials in the field of electric power engineering has allowed a fresh look at the problem of increasing economic efficiency, environmental performance and reliability of electric propulsion systems for different types of vehicle traction drives.

Considering the more demanding requirements regarding reliability and fault tolerance, efficiency, and the limitations on installation space and weight of traction drive components, the correct choice of topologies and design features of a vehicle traction drive is critical and must be based on a systematic approach.

The solution of this problem is especially important for Arctic icebreaker ships (AIS). That is because the correct choice of the optimal type of propulsion system for such ships is crucial for their successful operation in the difficult conditions of ice navigation.

Despite the relatively low capital cost of the ship's propulsion system (it can be from 15 to 30% of the total cost of the ship), the value of the operational costs associated with the propulsion system (maintenance, repairs, fuel consumption, ecological damage, etc.) can reach 90% of the ship's total operating costs.

I. Bolvashenkov et al., *Safety-Critical Electrical Drives*, SpringerBriefs in Electrical and Computer Engineering, https://doi.org/10.1007/978-3-319-89969-5_1

This chapter describes the experience of using stochastic models to assess the comprehensive operational efficiency of propulsion systems in icebreaker ships consisting of a few generating units and one or more traction motors. The Multi-Power Source Traction Drive (MPSTD) is the propulsion system that is typically used in icebreakers, Arctic cargo ships, diesel-electric locomotives and hybrid trucks. Its main purpose is to increase the fault tolerance and survivability of the vehicle, which are safety-critical systems.

The deterministic approach does not allow us to evaluate the probabilistic character of operational processes, such as the energy output of combustion engines, operational fuel consumption, load modes of traction electric motors or a large number of random factors affecting the accuracy of the simulation results.

The results of previous studies have proven the competitiveness of electric propulsion systems for certain types of ships, which are operated mostly under the partial power modes of the main traction system [2, 6]. Thus, the problem of determining the optimal type of propulsion system for icebreaker Arctic ships in terms of a systematic approach has become an urgent issue.

This approach to solving complex multi-criteria problems is particularly important in the analysis and evaluation of operational factors in Arctic icebreaker ships due to their highly probabilistic nature and immense significance. Therefore this chapter will focus on the choice of the optimal type of propulsion system for such ships.

Especially important here for the correct solution of this problem is the designing stage of Arctic icebreaker ships. In order to determine the most effective propulsion system for the transport ships navigating through ice, we developed a methodology of comparative analysis and stochastic models to estimate the values of specific operating parameters.

Until now, in their efforts to solve the problems involved in choosing the optimal type of propulsion system for cargo icebreaker ships for the Arctic region, designers at the various shipbuilding companies have used their intuition, based on their manufacturing experience and on traditional technological ideas [3]. Ships with different types of propulsion systems have been created for similar operations. For example, diesel-electric power drive (DED), which is based on a main electric motor and a fixed pitch propeller, and diesel-geared mechanical power drive (DGD), which is based on a medium rotational speed diesel motor, a reducer and a controllable pitch propeller have been developed for approximately the same ice conditions.

Each type of propulsion system has its own advantages and disadvantages. Therefore, the choice of an optimal system will have a decisive influence on the successful operation and technical and economic performance of an icebreaker ship. A detailed description of comparable propulsion systems for Arctic icebreaker ships is discussed in the next section.

1.2 A Comparative Analysis

This section provides an overview of two types of icebreaker ships for the purpose of a comparative analysis. Unlike traditional icebreakers, Arctic icebreaker ships (AIS) are multi-purpose vessels, designed to carry various cargoes under ice conditions. At the same time, they are also able to operate autonomously under light ice conditions or with the help of an icebreaker. Under light ice conditions, such ships can assist other ships that are not adapted for ice navigation.

The main purpose of Arctic icebreaker ships is to operate reliably and efficiently in the delivery of various cargoes in the Arctic area. For our comparative analysis, the Amguema-type Arctic icebreaker ship with DED and the Norilsk-type Arctic icebreaker ship with DGD were selected.

We decided to compare these two types of icebreaker ships for the following reasons:

- Both types are built for similar Arctic operating conditions.
- Both types of ship are operated by the same shipping company.
- Much statistical data has been collected and systematized during the five years of the operation of these two types of ship.
- Due to the considerable capital required and the operational costs and the immense cost of repairing damage to such vessels, an incorrect choice of the type of AIS traction drive, when calculated for a full life cycle, can lead to significant financial losses and to a significant decrease in the ship-owner's profits.

As already noted, each type of propulsion system for an icebreaker ship has its advantages and disadvantages. The quality indicators of each type of traction drive will depend significantly on the operating conditions. In order to investigate the impact of operating conditions on the efficiency of icebreaker ships with various traction drives, we developed a technique, described in Sect. 1.5.

The main advantages of ships with diesel-electric power drive [3, 4] are as follows:

- The ability of the primary diesel engine to operate in a constant and optimal mode, and as a consequence, the high economic and environmental value of such ships, which can perform with partial and variable operating loads.
- The high rates of vitality of the propulsion system and the ship as a whole, a factor that is especially important in its unique operating conditions.
- The flexible arrangement of the equipment of the electric propulsion system on board.
- The ship's high maneuverability.

The main advantages of ships with DGD are:

- Lower capital cost.
- Lower weight and reduced size.
- Higher efficiency in dealing with constant nominal loads.

Below is a short description of the two types of Arctic icebreaker ships that we compared.

1.2.1 Amguema-Type Diesel-Electric Ship

We analyzed a conventional diesel-electric power drive, used in Amguema-type arctic cargo ships (Fig. 1.1a), which are based on a hybrid-electric propulsion system. The structure of the ship's diesel-electric traction drive is shown in Fig. 1.1b. The power drive system consists of four diesel generators (DGs), a main switchboard (MSB), an electric converter (EC), an electric motor (EM), and a fixed pitch propeller (FPP).

The first DED Amguema-type ship was built at the Kherson Shipbuilding Factory (USSR) in 1962 and utilized a direct current electric propulsion system. Between 1962 and 1975, 17 vessels of this type were built. Two ships of this type are still in operation: the diesel-electric Yauza (built in 1974) and the diesel-electric Mikhail Somov (built in 1975).

The total power of the four diesel-generators is 5,500 kW. Depending on the ice conditions, the amount of cargo and other navigational conditions, the ship's propulsion system operates with a varying number of diesel-generators. In this way, it is able to attain the required value of power and, as a consequence, the ship has a

(a)

(b)

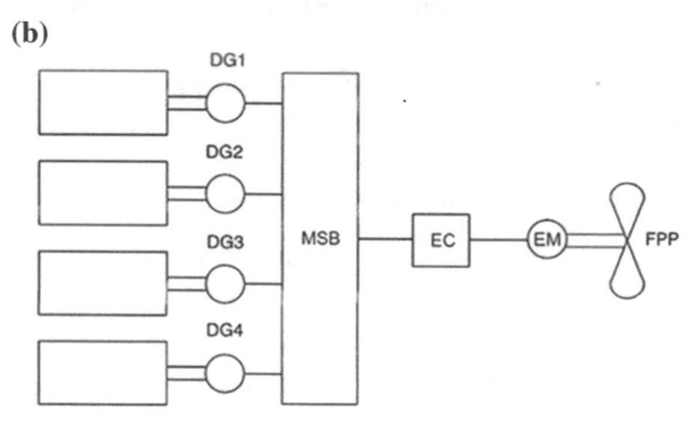

Fig. 1.1 **a** General view of an Amguema-type Arctic cargo ship, **b** structure of the diesel-electric drive of an Amguema-type arctic cargo ship

Table 1.1 Technical data on diesel-electric drive (DED)

Type of traction drive	DED
Main diesel engine (kW)	1,375 × 4
Rotational speed (1/min)	810
Electric propulsion motor	1
Maximum ice thickness in autonomous navigation (m)	0.8
Cargo capacity (tons)	7,900
Number of screws	1
Speed in ice-free water (knots)	14.3

high level of survivability of the possible critical failure of one or more of the traction drive's components. The maximum thickness of ice that an Amguema-type can overcome at a speed of 2 knots is 80 cm in a solid ice field (Table 1.1).

Each diesel-generator consists of a diesel engine and a generator. The power of each diesel-generator is 1,375 kW. Therefore, connecting diesel-generators in parallel support the nominal performance (power) required for the functioning of the whole system. During Arctic operations, the various schemes of multi-power sources can be applied. Depending on the ice navigational conditions, schemes with a varying number of diesel-generators can be used, in order to achieve an optimal steady operation state mode and optimal fuel consumption.

The main switchboard controls both the parallel operations of the diesel-generators and the power distribution between all the components of the propulsion system. The electric energy converter is intended to convert AC to DC.

The main direct current electric motor has a multi-drive design with a total power of two motor sections equal to 5,500 kW. In ice-free water operations, one electric motor section can be used.

In the ship's diesel-electric power drives with a fixed pitch propeller, the dimensioning of the electric machines has to be calculated accurately in order to estimate the available sufficient propulsion power, which is directly determined by the required value of operational power and additional required power in the case of heavy weather or ice conditions in the area of navigation.

The required structures of an arctic ship's propulsion system with a varying number of diesel-generators is determined by the operating requirements of the icebreaker ship as well as by ice, wind and temperature conditions.

1.2.2 Norilsk-Type Diesel-Geared Ship

As a comparative option, we analyzed the diesel-geared traction drive used in Norilsk-type Arctic icebreaker ships (Fig. 1.2a). The structure of the ship's diesel-geared traction drive is shown in Fig. 1.2b. The propulsion system consists

Fig. 1.2 **a** General view of a Norilsk-type diesel-geared ship, **b** structure of the diesel-geared drive of a Norilsk-type diesel-geared ship

(a)

(b)

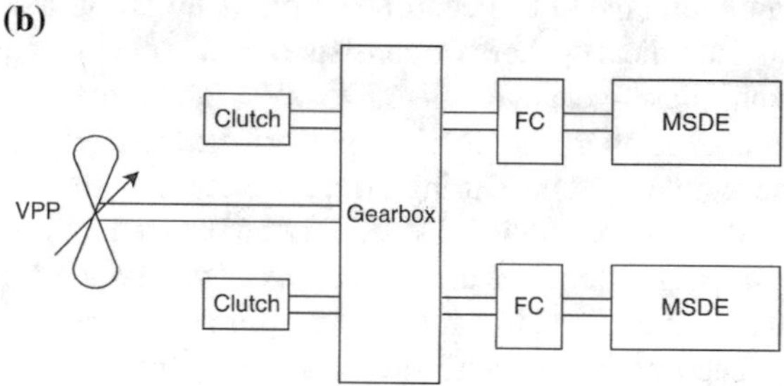

of two subsystems, each of which consists of a medium-speed diesel engine (MSDE), a fluid coupling (FC) and a clutch, a gearbox and a variable pitch propeller (VPP). Depending on operating conditions, the propulsion system can operate in one of two programs, namely, an ice program or a combined program.

The first Norilsk-type ship with DGD was built at the Wärtsilä Oy shipyard in Turku (Finland) in 1982 and is based on a diesel-geared mechanical propulsion system. Between 1982 and 1987 19 ships of this type were built. Two vessels of this type are still in operation: the diesel-geared Yuri Arshenevsky (built in 1986) and the diesel-geared Captain Danilkin (built in 1987).

The installed power of the whole generating system is 15,440 kW. Depending on the ice conditions, the amount of cargo and other conditions of navigation, the ship's propulsion system operates with both power generating subsystems or with only one of them. It can thus attain the required value of power performance and, as a consequence, the ship has a high level of survivability in the face of the possible critical failure of one or more of the propulsion system's components. The maximum thickness of ice that a Norilsk-type ship can overcome at a speed of 2 knots is 80 cm in a solid ice field (Table 1.2).

Each generating subsystem consists of a medium-speed diesel engine (MSDE), a fluid coupling (FC) and a clutch. The power of each subsystem is 7,720 kW.

Table 1.2 Technical data on diesel-geared drive (DGD)

Type of traction drive	DGD
Main diesel engine (kW)	7,720 × 2
Rotational speed (1/min)	560
Maximum ice thickness in autonomous navigation (m)	0.8
Reducer	1
Cargo capacity (tons)	15,000
Number of screws	1
Speed in ice-free water (knots)	18.1

Therefore, connecting subsystems in parallel support the nominal power, required for the functioning of the whole system under ice conditions. The variable pitch propeller (VPP) is designed to limit the changes of the MSDE load, which depends significantly on the operating conditions, and this limitation on the changes of the MSDE's load is accomplished by the altering of the pitch of the screw. The FCs are designed to absorb the impact of shock loads on the diesel engine when the screw interacts with ice.

The number of power generating subsystems in a diesel-geared propulsion system is determined by the operating conditions of the ship, as well as by ice, wind and temperature conditions.

1.3 Operational Conditions

Statistical operational data has been collected by the Murmansk Shipping Company for three Amguema-type vessels with DED and seven Norilsk-type vessels with DGD during five years of operations between 1982 and 1986, when they were employed under Arctic ice conditions. Despite the fact that the statistical operational data has been collected by only one shipping company, the operating conditions of the compared vessels were significantly different. This is because Amguema-type ships spend the main part of their operating time as supply trip vessels, while Norilsk-type ships are used as linear cargo ships on the Murmansk-Dudinka line.

Icebreaker ships are designed for operations in the Western part of the Russian Arctic, including the basins of the Barents and Kara Seas. The region of the ships' operations is presented as a gray area in Fig. 1.3.

The annual variations of ice thickness and temperature in the Arctic region of navigation are shown in Fig. 1.4 [6].

During the navigation period and depending on operating conditions, an icebreaker ship has four basic operating modes, corresponding to the number and power of the main engines.

Fig. 1.3 Area of operations

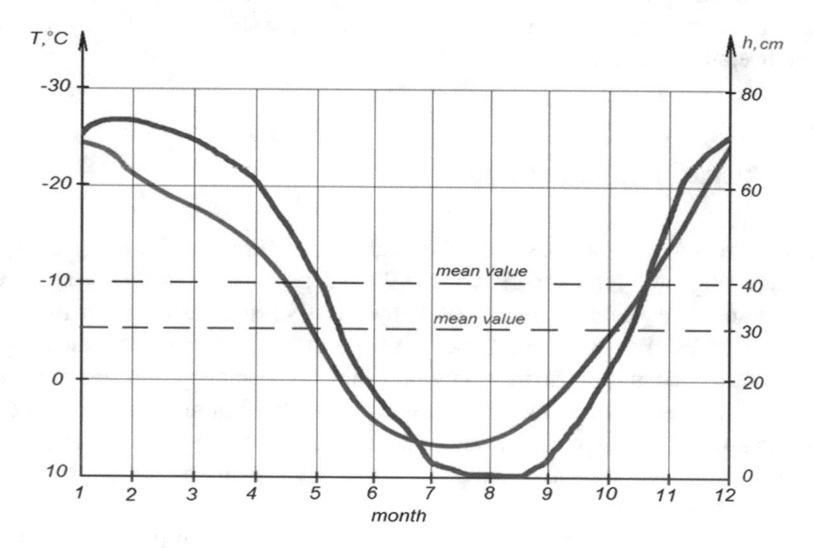

Fig. 1.4 Ice and temperature conditions of Western Arctic area

1.3.1 Navigation of a Ship in Ice-Free Water

The navigating mode of a vessel in ice-free water is characterized by a continuous use of the selected scheme and the power traction drive. In the overall balance of the running time of Arctic icebreaker ships and assuming that the vessel is used for its intended purpose during the whole operating period, this mode is employed approximately 15% of the time, as indicated by the statistical data collected by the Murmansk Shipping Company. The number and power of DGs and MSDEs will depend on what will ensure the sustainable motion of the ship with minimal fuel consumption.

1.3.2 Autonomous Movement in Ice Without a Convoy

The duration of this mode is approximately 44% of the navigation time of Arctic icebreaker ships. The choice of the operational scheme for the ship's propulsion system will depend on external (primarily ice value) conditions. In autonomous movement in the ice without a convoy, the scheme with two, three or four DGs on Amguema-type ships can be applied.

On Norilsk-type ships, depending on the condition and thickness of the ice that has to be overcome, it is possible to work with one or two main engines, in an ice or combined program. The working time of a scheme may be quite long and the power of the DGs and MSDEs will vary frequently.

1.3.3 Navigation of a Ship in a Convoy with an Icebreaker

An analysis of the statistical data indicates that this mode accounts for about 24% of ice navigation time. In this mode, on Amguema-type ships, a scheme with three or four DGs is usually applied, while, on Norilsk-type ships, the joint work of the two MSDEs will be used together with the ice or combined program.

The power of DGs and MSDEs will be used to ensure both the predetermined distance between the ships and the given speed of the whole convoy. The value of the velocity will depend on the ice conditions and the icebreaking capability of the least powerful ship in the convoy.

1.3.4 Maneuvering of the Ship

For Arctic icebreaker ships, maneuvering modes with frequent reverses of the propeller and with changes in the structure and power of the propulsion system constitute a large portion of the running time (about 17%). In this mode, the ship overcomes ice bridges, jams its body into the ice, tows stranded cargo ships that lack icebreaking capability, and operates in narrow places and in harbors.

The loads in these cases are unstable and variable over a wide range (from no load to maximum value). On Amguema-type ships, all possible schemes of propulsion system are used. On Norilsk-type ships, the separate or joint operating modes of the MSDE are used with two possible programs, ice or combined.

1.3.5 Required Power

According to the collected statistical data on the long-term operations of the ships under the ice conditions of the Arctic, the operational load modes of the DED of Amguema-type ships are as follows:

- Navigation with an ice breaker in heavy ice and navigation without an ice breaker in icefields requires 75–100% of nominal generated power.
- Navigation in open water, depending on required velocity, requires the greater part of operational time: 50–75% of nominal generated power.

The typical operational load modes of the DGD in Norilsk-type ships are as follows:

- Navigation with an ice breaker in heavy ice and navigation without an ice breaker in an icefield requires 100% of nominal generated power.
- Navigation in open water depending on required velocity requires 50–100% of nominal generated power.

1.4 Problem Formulation

The main disadvantage of existing methods of comparative efficiency assessment in choosing the optimal propulsion system for ships in Arctic operations is that universal criteria for comparison use purely technical, rather abstract parameters (such as the coefficient of technical level, technical excellence, etc.) or purely economic parameters (capital expenditures, specific costs, life cycle costs, etc.) without taking into account specific ice operational conditions.

The main features of the problem are presented in Fig. 1.5.

The problem is complicated because of the limited number of icebreaker ships, the conditions of operation in the Arctic region, the significant difference between these two types of ship in terms of design parameters and technological level, and such factors as the date and country of the ship's construction, and the shipbuilding company.

To evaluate the operational efficiency of the compared alternatives, it is necessary to create stochastic models of ships functioning under real Arctic ice conditions and to simulate compared variants under identical operating conditions. Thus, there

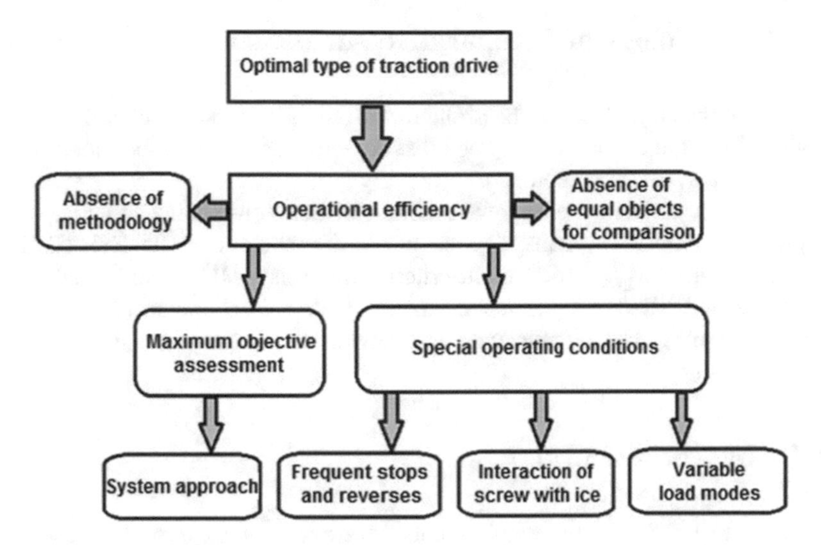

Fig. 1.5 Features of the problem

is a need for a statistical analysis of the operational modes of ships intended for the Arctic region in order to determine the real operating characteristics and to perform the systematization and formalization of statistical data for further use in the developed stochastic models of selected parameters for assessment. The results of the statistical analysis are given in Sect. 1.6.

The specific operating conditions of an icebreaker ship are as follows:

- dynamic loads of the interaction between the screw and the ice;
- frequent modes of variable loads;
- frequent stops-starts and reverses;
- low temperatures and strong winds;
- the significant influence of the human factor.

The main difficulties for carrying out a comparative analysis are:

- different periods of construction and, therefore, different levels of technological performance;
- significantly different technical data on the ships;
- lack of objective data on fuel consumption;
- different operational areas of the Arctic, hence significant differences in ice conditions of navigation;
- probabilistic nature of most of the variables;
- a number of uncertain factors.

These difficulties require a new approach for a comprehensive comparative assessment, whose main conditions are summarized in the following section.

1.5 Methodology of Comparative Assessment

Considering the complexity of the problem, its strongly marked probabilistic nature and a number of uncertain factors, the most objective decision can be found only on the basis of a systematic approach.

The study of different well-known methods of decision-making and comparative analysis of complex systems [7], such as life cycle cost analysis (LCCA), cost-benefit analysis (CBA), multi-criteria analysis (MCA) and analyses of advantages and disadvantages led us to create our own universal technique of comprehensive operational efficiency assessment for marine propulsion systems.

1.5.1 Generalized Criteria

According to [1, 5] and the requirements of a systematic approach, "sustainable delivery of the required amount of cargo under ice conditions" is accepted as the main target function for Arctic icebreaker ships.

In accordance with this main target function, all the various parameters can be divided into two groups: those relating to the direct attainment of the ship's target function, or the criteria of "usefulness," and the parameters related to operational costs and damage repair costs for the attainment of the target function, or the criteria of "payment for usefulness."

Thus, "usefulness" is the non-financial analog of the benefits criterion of CBA and "payment for usefulness" is the analog of the financial criterion of LCCA.

In accordance with the proposed technique and in consideration of the requirements and conditions for the operation of Arctic icebreaker ships, a ship's transportation performance A for its entire lifetime has been accepted as a criterion of "usefulness":

$$A = DVt_d. \tag{1.1}$$

$$t_d = t_o k_d. \tag{1.2}$$

where D is the amount of transported cargo in tons; V the ship's speed in knots, t_o and t_d are operational and driving time in hours respectively and k_d is the driving time rate.

For the second complex criterion—"payment for usefulness"—the sum C represents the capital, operating and damage costs for the entire lifetime of the ship:

$$C = C_{CAP} + C_{CONST} + C_F + C_{RAC}. \tag{1.3}$$

where C_{CAP} is capital costs, C_{CONST} fixed operating costs (personnel, navigation fees, taxis, insurance, etc.), C_F operating costs for fuel and oil, and C_{RAC} reliability-associated costs [8].

The values of the parameters in the two generalized criteria will depend on the operational conditions of the ship.

From the standpoint of operating conditions, the most informative parameters are the ship's speed, operating costs for fuel and oil and reliability-associated costs. These parameters define the values of a ship's operational performance, which will differ from the nominal values and will be largely determined by a number of operational factors (mode of generation and consumption of energy, fuel type, external conditions of navigation, etc.).

1.5.2 Local Criteria

Given the probabilistic nature of the ice navigation of an Arctic icebreaker ship, it was advisable to introduce the value of the operational speed of the ship in the form of a multi-factor regression model. As a result of the correlation analysis, the model of the ship's speed may be presented in the following form:

$$V = f(N, D, h, T_o, F, \beta). \tag{1.4}$$

where N is the power of the traction drive, D the amount of transported cargo, h the thickness of the ice, T_o the outside temperature, and F and β the strength and direction of the wind respectively. It should be noted that, in the regression model, the statistical value of the variable V takes into account not only the impact of the factors explicitly included in the model, but also other factors not included in the model.

It is especially important to emphasize that the regression model of operational speed, along with the above local parameters of "usefulness," allow an evaluation of the impact of non-metric indirect parameters, such as the type of ship body design, the propulsive quality of the ice, maneuverability and other unmeasurable factors.

Thus, for the calculation of the generalized criterion of "usefulness," it is necessary to determine the probability distribution (or average operational values) of these factors for the investigated operational area, which is discussed in the next section.

The calculation of the values of C_{CAP} and C_{CONST} is not a complicated problem and they do not depend on the operating conditions of the ship; thus, we will pay attention to the definition of C_F and C_{RAC}. Considering the statistical data on modern ships, fuel costs can account for up to 70% of total operating costs and will mostly depend on a ship's operating conditions.

The value of C_F was calculated based on the Markov model of energy generation by the diesel engines for the entire period of the ship's operation:

$$C_F = c_f \sum_j e_j g_j. \quad (1.5)$$

where c_f is specific fuel cost, e_j energy generation in the j-mode and g_j specific fuel consumption in the j-mode.

Based on statistical operational data, the value of the availability of the ship's propulsion system K_A for the ship's entire lifetime was defined. The value of the reliability-associated costs was calculated using the formula:

$$C_{RAC} = c_{UR} t_o k_d (1 - K_A). \quad (1.6)$$

where c_{UR} is the specific cost of ship downtime due to the propulsion system's unreliability and K_A the propulsion system's availability. The average cost of demurrage takes into account the full range of costs associated with the maintenance and repair of the propulsion system, activities that make it necessary to render the ship non-operational.

The solution to the problem of evaluating the reliability indices of the compared options for an Arctic icebreaker ship's propulsion systems is discussed in detail in Chap. 2.

1.6 Statistical Analysis

1.6.1 Multi-power Sources Energy Generation

To build a stochastic model of MPSTD energy generation, we relied on the operational statistics for five years of operation of Amguema-type ships with DED and Norilsk-type ships with DGD that were collected by the Murmansk Shipping Company.

Figure 1.6 shows the histograms of the power distribution of several schemes of DED in the operating modes, namely with one and four diesel generators.

As a result of the analysis, it was found that the most accurate power distribution of the ships can be described by the Weibull distribution, as can be seen from Fig. 1.7.

For ships with DGD, the statistical histograms of power distribution are presented in Fig. 1.8.

In contrast with DED, it is advisable to describe the power distribution of DGD using the Combined Law, representing the sum of exponential (for the small power range) and Weibull (for the large power range), as shown in Fig. 1.9.

The analysis of the operating modes of Arctic cargo ships allows us to determine the estimation average values of operational parameters, such as:

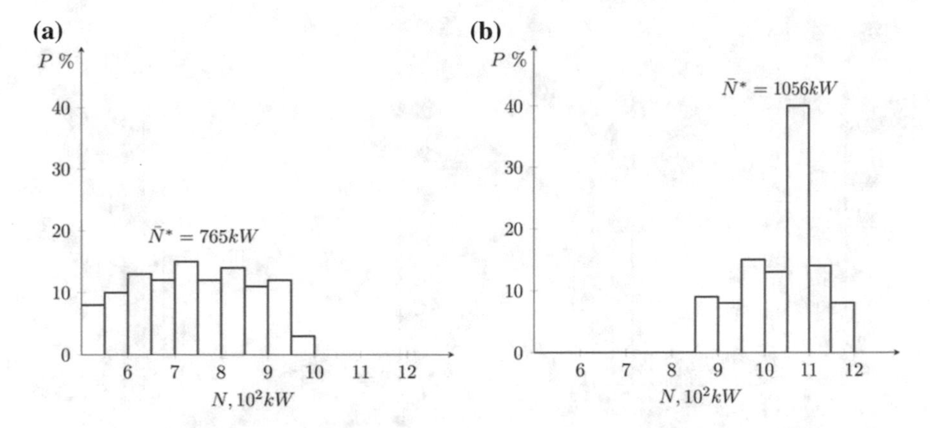

Fig. 1.6 Histograms of DED power distribution with one (**a**) and four (**b**) diesel-generators

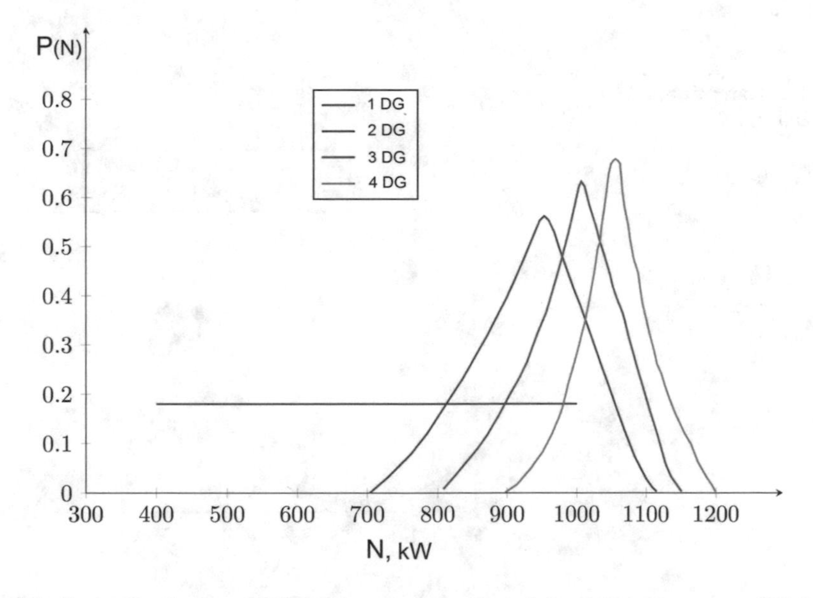

Fig. 1.7 Power distribution of DED with various numbers of diesel-generators: **a** one DG, **b** two DGs, **c** three DGs, **d** four DGs

- Operating time in Arctic ice navigation is at least 75% of total operating time.
- Running time is no less than 44% of the total operating time.
- The unavailability factor of DED is 0.01 and of DGD is 0.05.
- The ship's cargo capacity using ratio is 0.55.
- The average power using ratio of DED is 0.76 and of DGD is 0.56.
- The average speed using ratio of DED is 0.58 and of DGD is 0.53.

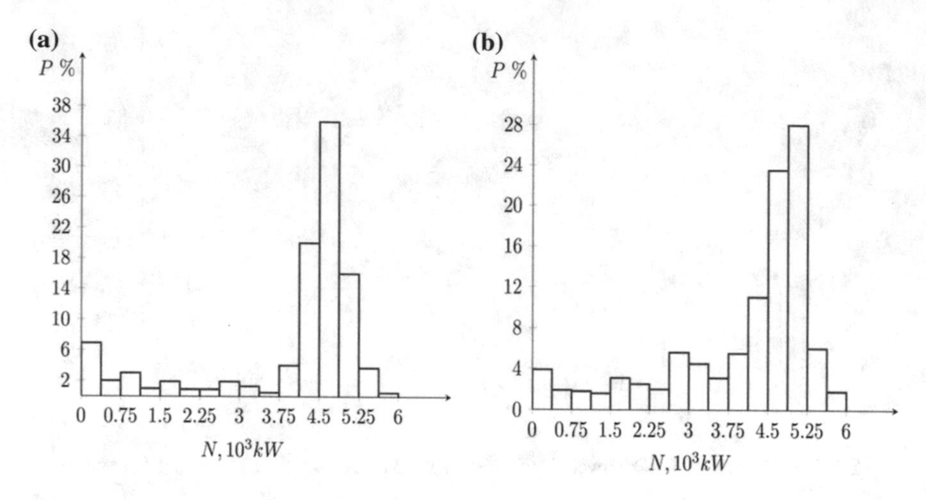

Fig. 1.8 Histograms of DGD power distribution: **a** single diesel operation, **b** double diesel operation

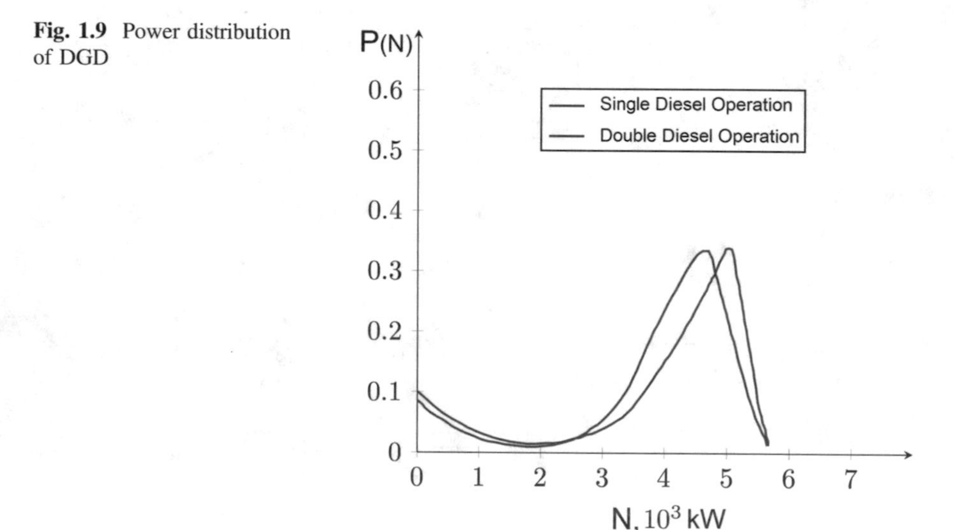

Fig. 1.9 Power distribution of DGD

1.6.2 *Specific Fuel Consumption*

It must be emphasized here that the nameplate data (nominal calculation) of the specific fuel consumption of the main diesel engines (for ships with DED-245 g/kWh and for ships with DGD-211 g/kWh) does not provide sufficiently informative indices for real Arctic ice conditions. Therefore we constructed the real ice operational characteristics of specific fuel consumption, dependent on the real power of the main engines, as shown in Figs. 1.10 and 1.11.

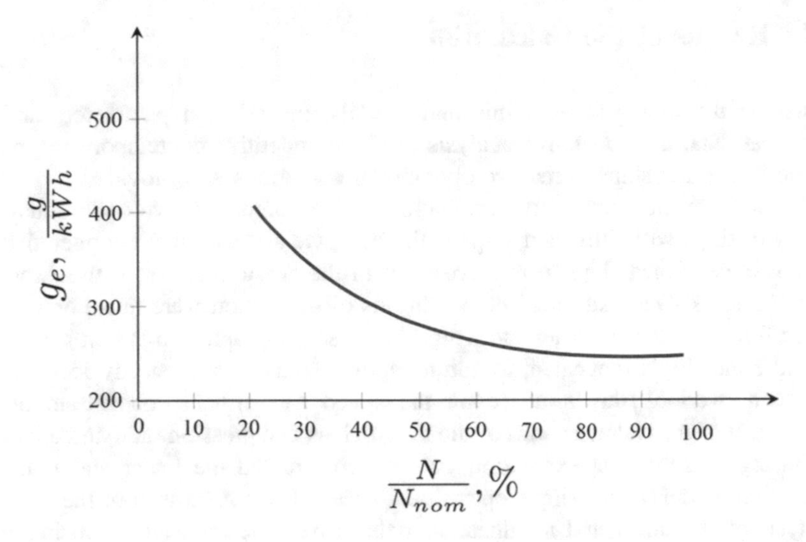

Fig. 1.10 Specific fuel consumption of the diesel engines of DED

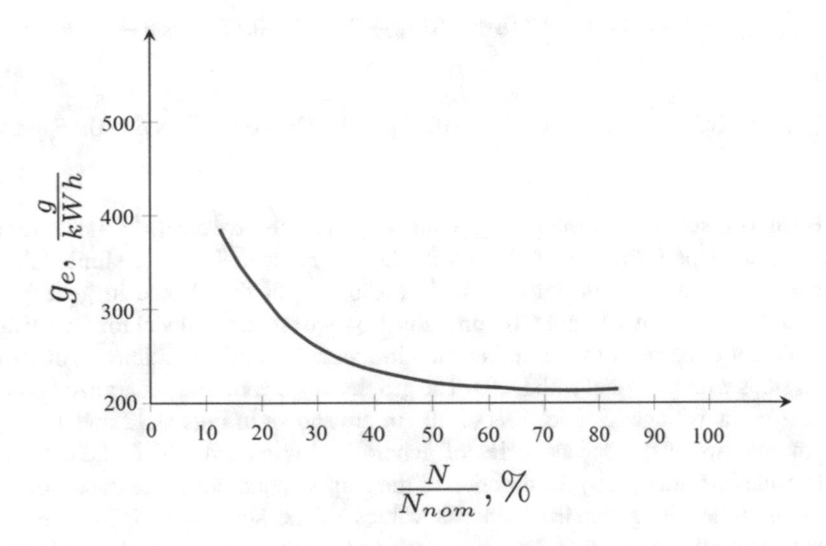

Fig. 1.11 Specific fuel consumption of the diesel engines of DGD

For the construction of these characteristics, 1-h measurements of fuel consumption and the corresponding values of main diesel power generation in various operating conditions were carried out. In addition, for the determination of the operational speed of the ship in real ice operations, the stochastic characteristics of the ice and meteorological conditions for a given area of operation were identified. As a result of the statistical data analysis, the expected value of average operational ice thickness that can be overcome by a ship without an icebreaker has been calculated. This value is approximately 30 cm.

1.7 Results of the Calculations

Based on the proposed technique and models, the selected parameters and the statistical data, a comparative analysis of the competitive contemporary types of Arctic icebreaker ships in real ice operational conditions was provided.

To perform the comparative evaluation of the comprehensive operational efficiency of ships with DED and ships with DGD, we assumed that the operations of the abovementioned ships were carried out in the Western sector of the Arctic on the Murmansk–Dudinka line. The conditions of navigation were thus comparable.

To formalize the character and features of a ship's propulsion systems, we must consider the ship's movement in autonomous navigation in a steady icefield. We can then overlook the influence on the speed and instead concentrate on the assessment of the characteristics of the ice, such as compression ratio and adhesion.

Basing ourselves on expression 1.4, we constructed the linear and nonlinear regression models of the ship's operational speed. The comparison of the accuracy analysis of the linear and nonlinear models proved the relevance in using models (1.7) and (1.8), as indicated below:

$$V_{DED} = -0.2 - 0.14D - 0.24h + 0.03T_o + 2.38N + 0.17F\cos\beta - 0.2F\sin\beta. \tag{1.7}$$

$$V_{DGD} = -30.4 - 0.03D - 0.11h + 0.14T_o + 0.97N + 0.08F\cos\beta - 0.1F\sin\beta. \tag{1.8}$$

Basing ourselves on the above regression dependences, we created a model for the transportation performance of ships with DED and with DGD in similar Arctic operational conditions. In order to study the nature of the change in generalized criteria A (expression 1.1) and C (expression 1.3), we created a model for their values for different durations of the running time in ice navigation conditions t_{io} of Arctic cargo ships with DED and with DGD. The graphs of the variation generalized criteria values for an average ice thickness of 30 cm are shown in Figs. 1.12 and 1.13.

An analysis of the dependencies of generalized criteria A and C indicates that their values are affected by the duration of the ship's operations in ice conditions. At the same time, the generalized criteria values of the ships with DGD are more dependent on the duration of the ice operations than the ships with DED, which are less dependent on ice navigation conditions.

For greater clarity and greater informational content, the comprehensive index Z was calculated using the following formula:

$$Z = (C_{CAP} + C_{CONST} + C_F + C_{RAC})/A. \tag{1.9}$$

The graphic dependencies of the comprehensive index Z on the relative operating time for diesel-electric drive and diesel-geared drive in various ice conditions are presented in Fig. 1.14.

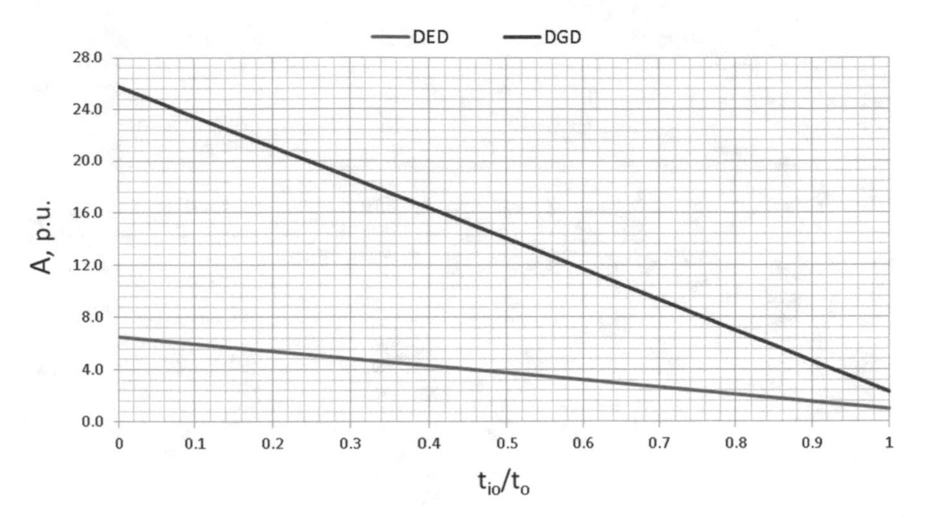

Fig. 1.12 Generalized criterion *A* (30 cm of ice thickness)

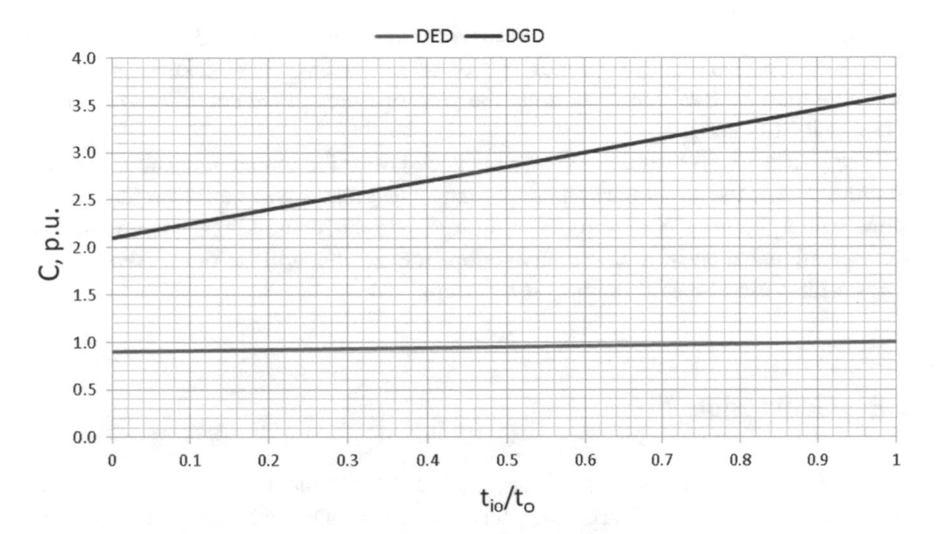

Fig. 1.13 Generalized criterion *C* (30 cm of ice thickness)

Based on the above graphics, it is possible to perform a comparative assessment of the comprehensive effectiveness of Arctic icebreaker ships with diesel-electric drive and with diesel-geared drive for various predetermined ice conditions (thickness of ice, temperature, wind, etc.) and for various durations of operational time in ice conditions.

We concluded that, under conditions of 30 cm average ice thickness, a ship with DED is more efficient, compared to a ship with DGD. The same conclusion may be made when, for more than 48% of the total operating time, the ship operates in ice

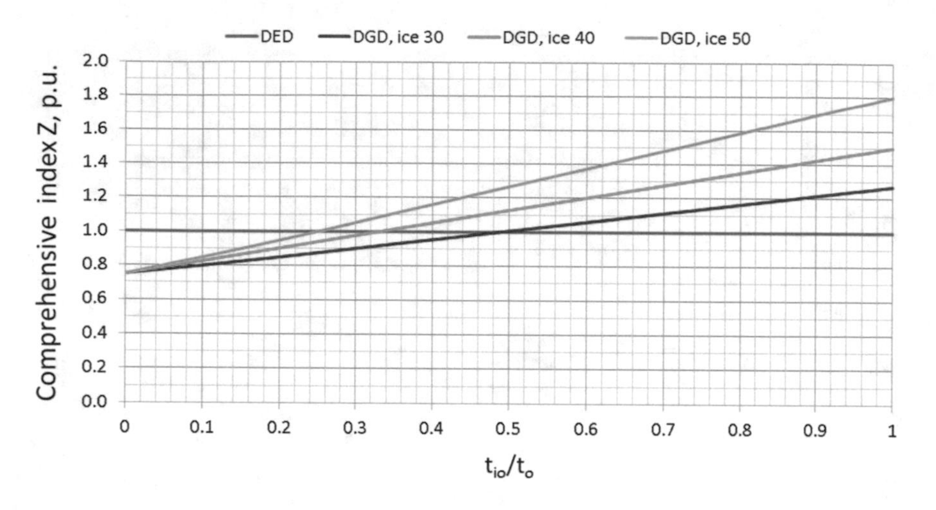

Fig. 1.14 Comprehensive index Z

conditions of 30 cm average ice thickness, or when, for more than 33% of operating time, the ship operates in ice conditions of 40 cm average ice thickness, or when, for more than 25% of operating time, the ship operates in ice conditions of 50 cm average ice thickness.

The results of the comparative estimates lead us to the main conclusion, namely, that the thicker the ice in the operational area, the greater will be the preference for ships with a diesel-electric propulsion system. Similarly, the longer the duration of the operational time of a ship's navigation, the greater will be the preference for ships with a diesel-electric propulsion system.

1.8 Conclusion

The methods of stochastic modeling, including statistical tests, regression and correlation analysis and Markov models, can solve the challenging task of conducting a comprehensive comparative assessment of the operational efficiency of vehicles with a similar functional orientation (in this case, Arctic icebreaker ships) but with a different design, different technical characteristics, and different conditions of ice navigation.

To simulate the operational modes of icebreaker ships in real conditions for the calculation of operational cost values of fuel, maintenance and repair, the most appropriate technique is to use Markov models. On the other hand, to simulate the criteria of the "usefulness" of icebreaker ships, the most convenient technique is to use multi-factor regression models.

Our proposed approach and methodology combine the advantages of LCC, CBA, MCA and the powerful tool of stochastic modeling. We can then consider not only the impact of the structural and functional features of a ship's propulsion system, but also the impact of specific conditions of its operation. The results we obtained are a good reason for objective decision-making regarding the choice of the optimal type of traction drive to be used in ice navigation ships, as well as for an analysis of possible activities and their potential for improving the efficiency of a ship's operation with various propulsion systems.

References

1. Bolvashenkov I, Herzog HG (2006) System approach to a choice of optimum factor of hybridization of the electric hybrid vehicle. In: Proceedings of electric vehicle symposium and exhibition (EVS), 23–28 Oct 2006, Yokohama, Japan, 2006, pp 1–12
2. Bolvashenkov I, Herzog HG (2016) Use of stochastic models for operational efficiency analysis of multi power source traction drives. In: Proceedings of the second international symposium on stochastic models in reliability engineering, life science and operations management, (SMRLO'16), 15–18 Feb 2016, Beer Sheva, Israel, 2016, pp 124–130
3. Bolvashenkov I, Herzog HG, Rubinraut A, Romanovskiy V (2014) Possible ways to improve the efficiency and competitiveness of modern ships with electric propulsion systems. In: Proceedings of 10th IEEE vehicle power and propulsion conference (VPPC'14), 27–30 Nov 2014, Coimbra, Portugal, 2014, pp 1–8
4. Bolvashenkov I, Shegalov I (1987) Gas turbine electric drive or diesel drive? Mar Fleet 10: 46–48 (in Russian)
5. Brahman T (1984) Multicriteriality and choice of the alternative in technique. Radio and Communications, Moscow (in Russian)
6. Frenkel I, Bolvashenkov I, Herzog HG, Khvatskin L (2016) Performance availability assessment of combined multi power source traction drive considering real operational conditions. Transp Telecommun 17(3):179–191
7. Frenkel I, Bolvashenkov I, Herzog HG, Khvatskin L (2017) Operational sustainability assessment of multi power source traction drive. In: Ram M, Davim JP (eds) Mathematics applied to engineering. Elsevier, London, pp 191–203
8. Lisnianski A, Frenkel I, Ding Y (2010) Multi-state system reliability analysis and optimization for engineers and industrial managers. Springer, London

Chapter 2
The Lz-Transform Method for the Reliability, Fault Tolerance and Operational Sustainability Assessment of Multi-power Source Traction Drives

Abstract This chapter focuses on a comparative analysis of the most important parameters of an icebreaker ship's sustainable operations: the operational availability, power performance and power performance deficiency of the multi-state Multi-Power Source Traction Drives of Amguema- and Norilsk-type Arctic icebreaker ships. These parameters, as was shown in Chap. 1, have a significant impact on the correct choice of the propulsive system of icebreaking vessels. The parameters' evaluation was based on statistical operational data on Arctic icebreaker ships with diesel-electric or diesel-geared traction drive. The *Lz*-transform approach was used to arrive at a solution of that problem. This approach drastically simplifies the solution compared with the straightforward Markov method.

Keywords Multi-power source traction drive · Icebreaker ship
Multi-state systems reliability · *Lz*-transform method · Availability
Sustainability · Power performance · Power performance deficiency

2.1 Introduction

Vehicle traction drives are safety-critical systems, which are subject to stringent requirements for safety, survivability and sustainability. Especially important is the implementation of these requirements for Arctic icebreaker ships. In this chapter, a comparative analysis of important parameters of an icebreaker ship's operational sustainability will be presented. Two Multi-Power Source Traction Drives of Arctic icebreaker ships will be analyzed: the diesel-electric traction drive of Amguema-type Arctic icebreaker ships and the diesel-geared traction drive of Norilsk-type Arctic icebreaker ships.

Due to the nature of a propulsion system, a fault in a single unit has only a partial effect on the entire power performance. A partial failure of the multi-power source traction drive initially and automatically leads to a partial system failure (reduction of output nominal power), as well as multiple consecutive failures and ultimately to

I. Bolvashenkov et al., *Safety-Critical Electrical Drives*, SpringerBriefs in Electrical and Computer Engineering, https://doi.org/10.1007/978-3-319-89969-5_2

a total system failure. Thus, a ship's multi-power source traction drive can be regarded as a multi-state system (MSS) whose components as well as the whole system can be considered to have a finite number of states associated with various performance rates [1–6]. The system's performance rate (output nominal power) can be viewed as a discrete-state continuous-time stochastic process. Such models, even in simple settings, are quite complex because they may contain several hundred states. Therefore, not only the construction of such a model but also the solution of the associated system of differential equations via a straightforward Markov method is very complicated.

In recent years a special technique known as *Lz*-transform has been proposed and investigated [7] for discrete-state continuous-time Markov processes. This approach is an extension of the universal generating function (UGF) technique proposed by Ushakov [9] that has been extensively implemented for the analysis of the reliability of multi-state systems. *Lz*-transform has turned out to be a powerful and highly efficient tool for the availability analysis of MSSs needed for constant and variable demand [3, 6, 10]. It should be noted here that the above technique has great applicability for numerous structure functions [8, 11].

In this chapter, the *Lz*-transform method is applied for the analysis of an MSS multi-power source traction drive that must function under various weather conditions: Its availability, power performance and power performance deficiency are investigated. We established that the implementation of the *Lz*-transform simplifies things considerably, as compared with the standard Markov model for the computation of a system's availability.

2.2 Brief Description of the *Lz*-Transform Method

Let us consider here a multi-state system, consisting of n multi-state components. Any j-component can have k_j different states, corresponding to different performances g_{ji}, represented by the set $\mathbf{g}_j = \{g_{j1}, \ldots, g_{jk_j}\}$, $j = \{1, \ldots, n\}$; $i = \{1, 2, \ldots, k_j\}$. The performance stochastic processes $G_j(t) \in \mathbf{g}_j$ and the system structure function $G(t) = f(G_1(t), \ldots, G_n(t))$ that produces the stochastic process corresponding to the output performance of the entire MSS fully define the MSS model.

The construction of MSS model definitions can be divided into several steps. For each multi-state component, we will build a model of stochastic process. The Markov performance stochastic process for each component j can be represented by the expression $G_j(t) = \{\mathbf{g}_j, \mathbf{A}_j, \mathbf{p}_{j0}\}$, where $\mathbf{g}_j$ is the set of possible component states, as defined below, $\mathbf{A}_j = \left(a_{lm}^{(j)}(t)\right)$, $l, m = 1, \ldots, k; j = 1, \ldots n$ the transition intensities matrix and $\mathbf{p}_{j0} = \left[p_{10}^{(j)} = \Pr\{G_j(0) = g_{10}\}, \ldots, p_{k_j0}^{(j)} = \Pr\{G_j(0) = g_{k_j0}\}\right]$ the probability distribution of initial states.

For each component j, the system of Kolmogorov forward differential equations [12] can be written for determination of state probabilities: $p_{ji}(t) = \Pr\{G_j(t) = g_{ji}\}$, $i = 1, \ldots k_j, j = 1, \ldots, n$ under initial conditions $\mathbf{p}_{j0}$. The *Lz*-transform of a discrete-state continuous-time (DSCT) Markov process $G_j(t)$ for each component j can be written as follows:

$$L_Z\{G_j(t)\} = \sum_{i=1}^{k_j} p_{ji}(t) z^{g_{ji}}. \tag{2.1}$$

The next step, which we must carry out in order to find the L_Z-transform of the MSS's entire output performance Markov Process $G(t)$ is to apply the Ushakov's Universal Generating Operator [9] to all individual *Lz*-transforms $L_Z\{G_j(t)\}$ over all time points $t \geq 0$:

$$L_Z\{G(t)\} = \Omega_f\{L_Z[G_1(t)], \ldots, L_Z[G_n(t)]\} = \sum_{i=1}^{K} p_i(t) z^{g_i} \tag{2.2}$$

The technique of Ushakov's operator application is well established for many different structure functions [8].

Using the resulting *Lz*-transform, MSS mean instantaneous availability for constant demand level w can be derived as the sum of all probabilities in *Lz*-transform for terms where g_i, the powers of z, are not less than demand level w:

$$A(t) = \sum_{g_i \geq w} p_i(t). \tag{2.3}$$

An MSS's mean instantaneous performance may be calculated as the sum of all probabilities multiplied to performance in *Lz*-transform for terms where the powers of z are positive:

$$E(t) = \sum_{g_i > 0} p_i(t) g_i. \tag{2.4}$$

The instantaneous performance deficiency $D(t)$ at any time t for constant demand w can be calculated as follows:

$$D(t) = \sum_{i=1}^{K} p_i(t) \cdot \max(w - g_i, 0). \tag{2.5}$$

2.3 Multi-state Models of Multi-power Source Traction Drives

2.3.1 Description of Systems

We can analyze a diesel-electric traction drive, used in Amguema-type Arctic cargo ships, by comparing it to a diesel-geared power drive, used in Norilsk-type Arctic cargo ships, which are based on a diesel-geared propulsion system.

2.3.1.1 The Amguema-Type Ship's Diesel-Electric Traction Drive

The structure of an Amguema-type Arctic cargo ship's diesel-electric traction drive is shown in Fig. 2.1. The system consists of a diesel generator subsystem, a main switchboard, an electric energy converter, an electric motor subsystem and a fixed pitch propeller.

The power performance of the whole system is 5,500 kW. Depending on ice conditions, the amount of cargo and other conditions of navigation, the ship's propulsion system operates with a varying number of diesel and electric propulsion motors. It realizes the required value of performance and as a consequence attains the high survivability of the ship in the face of the possible occurrence of the critical failure of the power equipment.

Each diesel generator consists of a diesel engine and a generator. The generated power performance of each diesel generator is 1,375 kW. The connected diesel generators in parallel support the nominal performance required for the functioning of the whole system.

The main switchboard device and electric energy converter have nominal performance.

The electric motor subsystem consists of two motors with a total performance is 2,750 kW each. The connected motors in parallel support the nominal performance required for the functioning of the whole system.

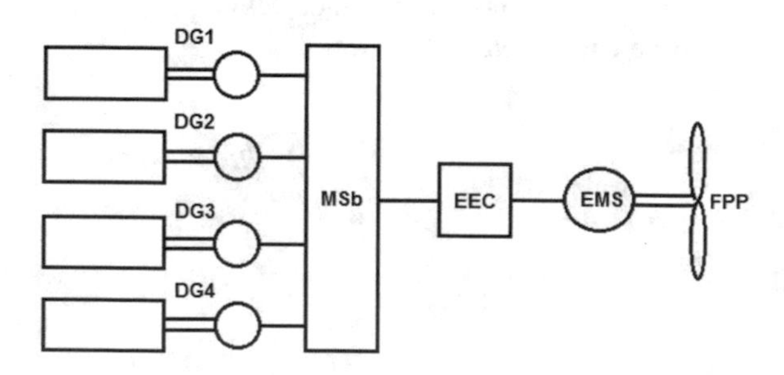

Fig. 2.1 Structure of an Amguema-type ship's diesel-electric traction drive

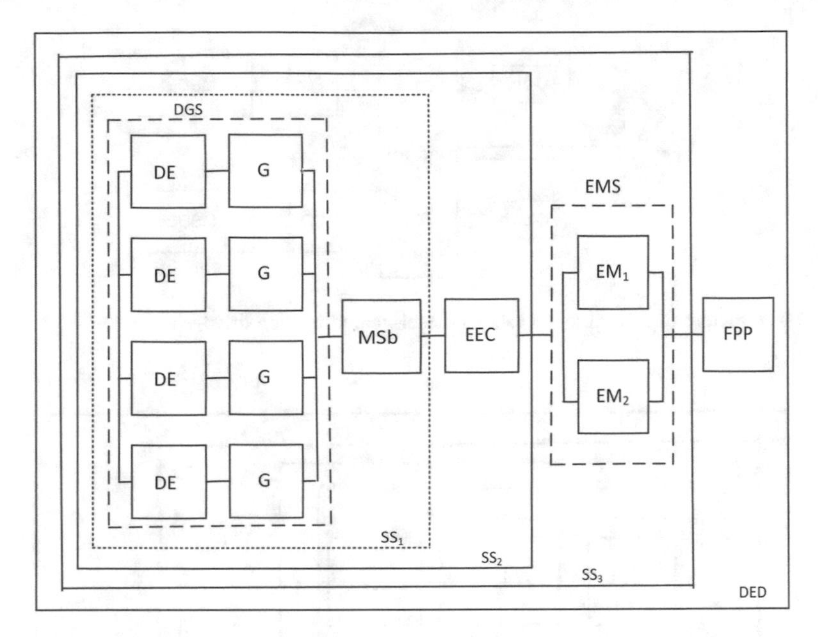

Fig. 2.2 Reliability block diagram of the diesel-electric traction drive in an Amguema-type ship

In diesel-electric power drives with a fixed pitch propeller, the dimensioning of the electric machines has to be calculated accurately in order to estimate both the available sufficient propulsion power, which is directly determined by the required value of operational power, and the additional power required in case of heavy weather or ice conditions in the area of navigation. The Reliability Block Diagram of the diesel-electric traction drive in an Amguema-type ship is presented in Fig. 2.2.

2.3.1.2 The Norilsk-Type Ship's Diesel-Geared Traction Drive

Diesel-geared power drive, used in Norilsk-type Arctic cargo ships, is based on a diesel-geared propulsion system. The structure of the ship's diesel-geared traction drive is shown in Fig. 2.3. The system consists of two subsystems, each consisting of a medium-speed diesel engine, a fluid coupling and a clutch, a gear box and a variable pitch propeller.

The power performance of the whole system is 15,440 kW. Depending on ice conditions, the amount of cargo and other conditions of navigation, the ship's propulsion system operates with both subsystems or with only one of them. It realizes the required value of power performance and as a consequence attains the high survivability of the ship in the face of the possible occurrence of the critical failure of the power equipment.

Each subsystem consists of a medium-speed diesel engine, a fluid coupling and a clutch. The power performance of each subsystem is 7,720 kW. The connected

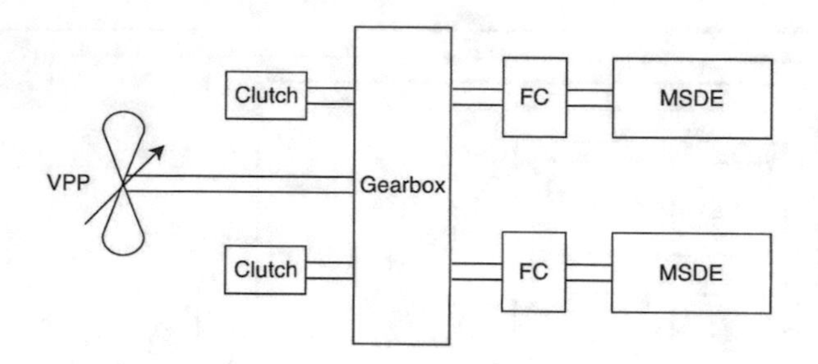

Fig. 2.3 Structure of a Norilsk-type ship's diesel-geared traction drive

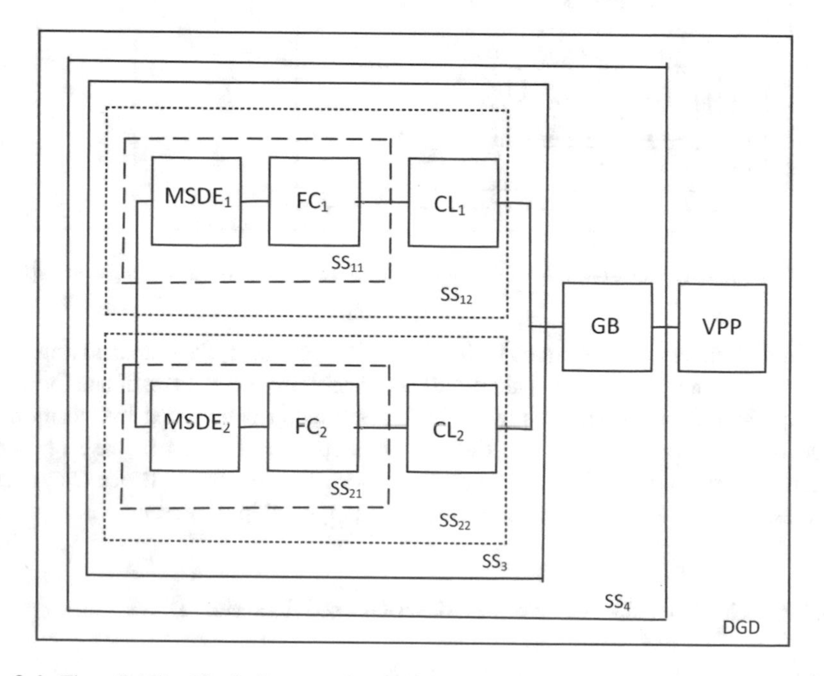

Fig. 2.4 The reliability block diagram of a Norilsk-type ship's diesel-geared traction drive

subsystems in parallel support the nominal performance required for the functioning of the whole system.

The gear box and variable pitch propeller have nominal performance.

As was pointed out in Chap. 1, the variable pitch propeller (VPP) is designed to limit the changes of the MSDE load, which depends significantly on the operating conditions. This limitation of the changes of the MSDE load is performed through changes in the pitch of the screw. The nominal power of the variable pitch propeller is the same as the nominal power of the whole system and totals 15,440 kW.

The Reliability Block Diagram of a Norilsk-type ship's diesel-geared traction drive is presented in Fig. 2.4.

The possible structures of an Arctic ship's propulsion system are determined by the operating conditions of the Arctic ship and by ice and temperature conditions.

The typical operational modes of Arctic cargo ships are as follows:

- Navigation with an ice-breaker in heavy ice and navigation without an ice-breaker in solid ice uses 75% of the generated power.
- Navigation in open water, depending on required velocity, uses 50% of the generated power.

2.3.2 Description of the System's Elements

The system's elements have two states (fully working and fully failed). According to *Lz*-transform method, described in the Sect. 2.2, in order to calculate the probabilities for each state, we built a state space diagram (Fig. 2.5) and the following system of differential equations:

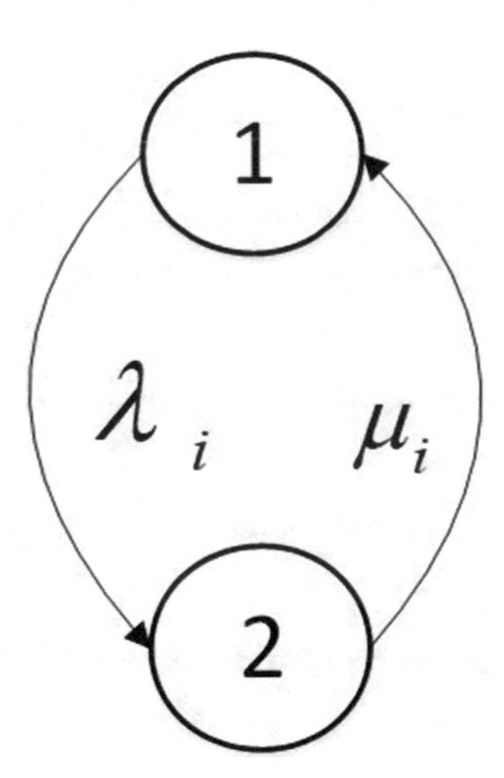

Fig. 2.5 State space diagram of the system's elements

$$\begin{cases} \dfrac{dp_{i1}(t)}{dt} = -\lambda_i p_{i1}(t) + \mu_i p_{i2}(t), \\ \dfrac{dp_{i2}(t)}{dt} = \lambda_i p_{i1}(t) - \mu_i p_{i2}(t). \end{cases}$$

i = DE, G, MSb, EEC, EM_1, EM_2 FPP, $MSDE_1$, $MSDE_2$, FC_1, FC_2, CL_1, CL_2, GB, VPP

Initial conditions are: $p_{i1}(0) = 1$; $p_{i2}(0) = 0$.

We used MATLAB® for the numerical solution of these systems of DE to obtain probabilities $p_{i1}(t), p_{i2}(t)$, (i = DE, G, MSb, EEC, EM_1, EM_2, FPP, $MSDE_1$, $MSDE_2$, FC_1, FC_2, CL_1, CL_2, GB, VPP).

Therefore, for the elements of such systems, the output performance stochastic processes can be obtained in the following manner:

- In an Amguema-type ship

For i = DE, G	For i = EM_1, EM_2	For i = MSb, EEC, FPP
$\begin{cases} \mathbf{g}_i = \{g_{i1}, g_{i1}\} = \{1,375, 0\} \\ \mathbf{p}_i(t) = \{p_{i1}(t), p_{i1}(t)\}. \end{cases}$	$\begin{cases} \mathbf{g}_i = \{g_{i1}, g_{i1}\} = \{2,750, 0\} \\ \mathbf{p}_i(t) = \{p_{i1}(t), p_{i1}(t)\}. \end{cases}$	$\begin{cases} \mathbf{g}_i = \{g_{i1}, g_{i1}\} = \{5,500, 0\} \\ \mathbf{p}_i(t) = \{p_{i1}(t), p_{i1}(t)\}. \end{cases}$

- In a Norilsk-type ship

For i = $MSDE_1$, $MSDE_2$, FC_1, FC_2, CL_1, CL_2	For i = GB, VPP
$\begin{cases} \mathbf{g}_i - \{g_{i1}, g_{i1}\} = \{7,720, 0\} \\ \mathbf{p}_i(t) = \{p_{i1}(t), p_{i1}(t)\}. \end{cases}$	$\begin{cases} \mathbf{g}_i = \{g_{i1}, g_{i1}\} = \{15,540, 0\} \\ \mathbf{p}_i(t) = \{p_{i1}(t), p_{i1}(t)\}. \end{cases}$

Sets $\mathbf{g}_i, \mathbf{p}_i(t)$ define Lz-transforms for each element in an Amguema-type ship as follows:

Diesel engine:

$$L_z\{g^{DE}(t)\} = p_1^{DE}(t)z^{g_1^{DE}} + p_2^{DE}(t)z^{g_2^{DE}} = p_1^{DE}(t)z^{1,375} + p_2^{DE}(t)z^0. \tag{2.6}$$

Generator:

$$L_z\{g^G(t)\} = p_1^G(t)z^{g_1^G} + p_2^G(t)z^{g_2^G} = p_1^G(t)z^{1,375} + p_2^G(t)z^0. \tag{2.7}$$

Main switchboard:

$$L_z\{g^{MSb}(t)\} = p_1^{MSb}(t)z^{g_1^{MSb}} + p_2^{MSb}(t)z^{g_2^{MSb}} = p_1^{MSb}(t)z^{5,500} + p_2^{MSb}(t)z^0. \tag{2.8}$$

Electric energy converter:

$$L_z\{g^{EEC}(t)\} = p_1^{EEC}(t)z^{g_{i1}^{EEC}} + p_2^{EEC}(t)z^{g_{i2}^{EEC}} = p_1^{EEC}(t)z^{5,500} + p_2^{EEC}(t)z^0. \tag{2.9}$$

Electric motor i = 1, 2:

$$L_z\{g^{EM_i}(t)\} = p_1^{EM_i}(t)z^{g_{i1}^{EM_i}} + p_2^{EM_i}(t)z^{g_{i2}^{EM_i}} = p_1^{EM_i}(t)z^{2,750} + p_2^{EM_i}(t)z^0. \tag{2.10}$$

Sets $\mathbf{g}_i, \mathbf{p}_i(t)$ define Lz-transforms for each element in a Norilsk-type ship as follows:

Medium Speed Diesel Engine:

$$\begin{aligned} L_z\{G^{MSDE_i}(t)\} &= p_1^{MSDE_i}(t)z^{g_1^{MSDE_i}} + p_2^{MSDE_i}(t)z^{g_2^{MSDE_i}} \\ &= p_1^{MSDE_i}(t)z^{7,720} + p_2^{MSDE_i}(t)z^0, \quad i = 1,2 \end{aligned} \tag{2.11}$$

Fluid coupling:

$$\begin{aligned} L_z\{G^{FC_i}(t)\} &= p_1^{FC_i}(t)z^{g_1^{FC_i}} + p_2^{FC_i}(t)z^{g_2^{FC_i}} \\ &= p_1^{FC_i}(t)z^{7720} + p_2^{FC_i}(t)z^0, \quad i = 1,2 \end{aligned} \tag{2.12}$$

Clutch:

$$\begin{aligned} L_z\{G^{CL_i}(t)\} &= p_1^{CL_i}(t)z^{g_1^{CL_i}} + p_2^{CL_i}(t)z^{g_2^{CL_i}} \\ &= p_1^{CL_i}(t)z^{7720} + p_2^{CL_i}(t)z^0, \quad i = 1,2 \end{aligned} \tag{2.13}$$

Gear box:

$$L_z\{G^{GB}(t)\} = p_1^{GB}(t)z^{g_1^{GB}} + p_2^{GB}(t)z^{g_2^{GB}} = p_1^{GB}(t)z^{15,440} + p_2^{GB}(t)z^0. \tag{2.14}$$

Variable pitch propeller:

$$L_z\{G^{VPP}(t)\} = p_1^{VPP}(t)z^{g_1^{VPP}} + p_2^{VPP}(t)z^{g_2^{VPP}} = p_1^{VPP}(t)z^{15,440} + p_2^{VPP}(t)z^0. \tag{2.15}$$

2.3.3 *Multi-state Model for the Amguema-Type Ship's Diesel-Electric Traction Drive*

As one can see in Fig. 2.2, the multi-state model for Multi-Power Source Traction Drive may be presented as connected in a series: diesel-generator sub-system, main switchboard, electric energy converter, electric motor subsystem and fixed pitch propeller. Therefore, the whole system L_Z-transform is as follows:

$$L_z\{G^{DED}(t)\} = \Omega_{f_{ser}}\left(L_z\{G^{DGS}(t)\}, L_z\{G^{MSb}(t)\}, L_z\{G^{EEC}(t)\}, L_z\{G^{EMS}(t)\}, L_z\{G^{FPP}(t)\}\right) \tag{2.16}$$

Using the recursive derivation approach [8], we can calculate the whole system *Lz*-transform as follows:

$$\begin{aligned} L_z\{G^{SS_1}(t)\} &= \Omega_{f_{ser}}\left(L_z\{G^{DGS}(t)\}, L_z\{G^{MSb}(t)\}\right) \\ L_z\{G^{SS_2}(t)\} &= \Omega_{f_{ser}}\left(L_z\{G^{SS_1}(t)\}, L_z\{G^{EEC}(t)\}\right) \\ L_z\{G^{SS_3}(t)\} &= \Omega_{f_{ser}}\left(L_z\{G^{SS_2}(t)\}, L_z\{G^{EMS}(t)\}\right) \\ L_z\{G^{DED}(t)\} &= \Omega_{f_{ser}}\left(L_z\{G^{SS_3}(t)\}, L_z\{G^{FPP}(t)\}\right). \end{aligned} \tag{2.17}$$

2.3.3.1 Diesel Generator Sub-system

The diesel generator sub-system consists of four pairs of identical diesel engines and generators connected in parallel. Each diesel engine and each generator are two-state devices: The performance of a fully operational state is 1,375 kW and total failure corresponds to a capacity of 0.

Using the composition operator Ω_{fser}, we obtain the L_Z-transform $L_z\{G^{DG}(t)\}$ for each pair of identical diesel engines and generators, connected in series, where the powers of z are calculated as the minimum of powers of the corresponding terms:

$$\begin{aligned} L_z\{G^{DG}(t)\} &= \Omega_{f_{ser}}\left(g^{DE}(t), g^{G}(t)\right) \\ &= p_1^{DE}(t)p_1^{G}(t)z^{1,375} + \left(p_1^{DE}(t)p_2^{G}(t) + p_2^{DE}(t)\right)z^0. \end{aligned} \tag{2.18}$$

Using the notations

$$\begin{aligned} p_1^{DG}(t) &= p_1^{DE}(t)p_1^{G}(t); \\ p_2^{DG}(t) &= p_1^{DE}(t)p_2^{G}(t) + p_2^{DE}(t); \end{aligned}$$

we obtain the resulting *Lz*-transform for the diesel generator sub-system:

$$L_z\{G^{DG}(t)\} = p_1^{DG}(t)z^{1,375} + p_2^{DG}(t)z^0. \tag{2.19}$$

Using the composition operator Ω_{fpar} for four diesel generators, connected in parallel, we obtain the L_Z-transform $L_z\{G^{DGS}(t)\}$ for the whole diesel generator sub-system:

$$L_z\{G^{DGS}(t)\} = \Omega_{f_{par}}\left(L_z\{G^{DG}(t)\}, L_z\{G^{DG}(t)\}, L_z\{G^{DG}(t)\}, L_z\{G^{DG}(t)\}\right). \tag{2.20}$$

Using notations

$$\begin{aligned}
&P_1^{DGS}(t) = \{p_1^{DG}(t)\}^4;\\
&P_2^{DGS}(t) = 4 \cdot \{p_1^{DG}(t)\}^3 p_2^{DG}(t);\\
&P_3^{DGS}(t) = 6 \cdot \{p_1^{DG}(t)\}^2 \{p_2^{DG}(t)\}^2;\\
&P_4^{DGS}(t) = 4 \cdot p_1^{DG}(t) \{p_2^{DG}(t)\}^3;\\
&P_5^{DGS}(t) = \{p_2^{DG}(t)\}^4.
\end{aligned}$$

we obtain the resulting L_Z-transform for the whole diesel generator sub-system:

$$\begin{aligned}
L_z\{G^{DGS}(t)\} = &P_1^{DGS}(t)z^{5,500} + P_2^{DGS}(t)z^{4,125} + P_3^{DGS}(t)z^{2,750}\\
&+ P_4^{DGS}(t)z^{1,375} + P_4^{DGS}(t)z^0.
\end{aligned} \tag{2.21}$$

2.3.3.2 Electric Motor Sub-system

The electric motor sub-system consists of two electric motors connected in parallel. Each electric motor is a two-state device: the performance of a fully operational state is 2,750 kW and total failure corresponds to a capacity of 0.

Using the composition operator Ω_{fpar} for two electric motors connected in parallel, we obtain the *Lz*-transform $L_z\{G^{EMS}(t)\}$ for the whole electric motor sub-system:

$$L_z\{G^{EMS}(t)\} = \Omega_{f_{par}}\left(L_z\{G^{EM_1}(t)\}, L_z\{G^{EM_2}(t)\}\right). \tag{2.22}$$

Using notations

$$\begin{aligned}
&P_1^{EMS}(t) = p_1^{EM_1}(t)p_2^{EM_2}(t);\\
&P_2^{EMS}(t) = p_1^{EM_1}(t)p_2^{EM_2}(t) + p_2^{EM_1}(t)p_1^{EM_2}(t);\\
&P_3^{EMS}(t) = p_2^{EM_1}(t)p_2^{EM_2}(t);
\end{aligned}$$

we obtain the resulting *Lz*-transform for the whole diesel generator sub-system:

$$L_z\{G^{EMS}(t)\} = P_1^{EMS}(t)z^{5,500} + P_2^{EMS}(t)z^{2,750} + P_3^{EMS}(t)z^0. \tag{2.23}$$

2.3.3.3 An *Lz*-Transform Calculation of a Diesel-Electric Propulsion System

Using the composition operator Ω_{fser} for sub-systems and elements connected in series, where the powers of z are calculated as the minimum values of the powers of corresponding terms, we obtain the following L_Z-transforms:

- L_Z-transforms for the SS_1 subsystem

$$\begin{aligned} L_z\{G^{SS_1}(t)\} &= \Omega_{f_{ser}}\left(L_z\{G^{DGS}(t)\}, L_z\{G^{MSb}(t)\}\right) \\ &= \Omega_{f_{ser}}\left(P_1^{DGS}(t)z^{5,500} + P_2^{DGS}(t)z^{4,125} + P_3^{DGS}(t)z^{2,750}\right. \\ &\quad \left. + P_4^{DGS}(t)z^{1,375} + P_5^{DGS}(t)z^0, p_1^{MSb}(t)z^{5,500} + p_2^{MSb}(t)z^0\right) \\ &= P_1^{SS_1}(t)z^{5,500} + P_2^{SS_1}(t)z^{4,125} + P_3^{SS_1}(t)z^{2,750} + P_4^{SS_1}(t)z^{1,375} + P_5^{SS_1}(t)z^0. \end{aligned} \tag{2.24}$$

where

$$\begin{aligned} P_1^{SS_1}(t) &= P_1^{DGS}(t)p_1^{MSb}(t), \\ P_2^{SS_1}(t) &= P_2^{DGS}(t)p_1^{MSb}(t), \\ P_3^{SS_1}(t) &= P_3^{DGS}(t)p_1^{MSb}(t), \\ P_4^{SS_1}(t) &= P_4^{DGS}(t)p_1^{MSb}(t), \\ P_5^{SS_1}(t) &= P_5^{DGS}(t)p_1^{MSb}(t) + p_2^{MSb}(t). \end{aligned}$$

- L_Z-transforms for the SS_2 subsystem

$$\begin{aligned} L_z\{G^{SS_2}(t)\} &= \Omega_{f_{ser}}\left(L_z\{G^{SS_1}(t)\}, L_z\{G^{EEC}(t)\}\right) \\ &= \Omega_{f_{ser}}\left(P_1^{SS_1}(t)z^{5,500} + P_2^{SS_1}(t)z^{4,125} + P_3^{SS_1}(t)z^{2,750}\right. \\ &\quad \left. + P_4^{SS_1}(t)z^{1,375} + P_5^{SS_1}(t)z^0, p_1^{EEC}(t)z^{5,500} + p_2^{EEC}(t)z^0\right) \\ &= P_1^{SS_2}(t)z^{5,500} + P_2^{SS_2}(t)z^{4,125} + P_3^{SS_2}(t)z^{2,750} + P_4^{SS_2}(t)z^{1,375} + P_5^{SS_2}(t)z^0. \end{aligned} \tag{2.25}$$

where

$$
\begin{aligned}
P_1^{SS_2}(t) &= P_1^{SS_1}(t)p_1^{EEC}(t),\\
P_2^{SS_2}(t) &= P_2^{SS_1}(t)p_1^{EEC}(t),\\
P_3^{SS_2}(t) &= P_3^{SS_1}(t)p_1^{EEC}(t),\\
P_4^{SS_2}(t) &= P_4^{SS_1}(t)p_1^{EEC}(t),\\
P_5^{SS_2}(t) &= P_5^{SS_1}(t)p_1^{EEC}(t) + p_2^{EEC}(t).
\end{aligned}
$$

- L_Z-transforms for the SS_3 subsystem

$$
\begin{aligned}
L_z\{G^{SS_3}(t)\} &= \Omega_{f_{ser}}\left(L_z\{G^{SS_2}(t)\}, L_z\{G^{EMS}(t)\}\right)\\
&= \Omega_{f_{ser}}\left(P_1^{SS_2}(t)z^{5{,}500} + P_2^{SS_2}(t)z^{4{,}125} + P_3^{SS_2}(t)z^{2{,}750}\right.\\
&\quad + P_4^{SS_2}(t)z^{1{,}375} + P_5^{SS_2}(t)z^0, p_1^{EMS}(t)z^{5{,}500}\\
&\quad \left. + p_2^{EMS}(t)z^{2{,}750} + p_3^{EMS}(t)z^0\right)\\
&= P_1^{SS_3}(t)z^{5{,}500} + P_2^{SS_3}(t)z^{4{,}125} + P_3^{SS_3}(t)z^{2{,}750}\\
&\quad + P_4^{SS_3}(t)z^{1{,}375} + P_5^{SS_3}(t)z^0.
\end{aligned}
\tag{2.26}
$$

where

$$
\begin{aligned}
P_1^{SS_3}(t) &= P_1^{SS_2}(t)P_1^{EMS}(t),\\
P_2^{SS_3}(t) &= P_2^{SS_2}(t)P_1^{EMS}(t),\\
P_3^{SS_3}(t) &= P_1^{SS_2}(t)P_2^{EM}(t) + P_2^{SS_2}(t)P_2^{EMS}(t)\\
&\quad + P_3^{SS_2}(t)P_1^{EMS}(t) + P_3^{SS_2}(t)P_2^{EMS}(t),\\
P_4^{SS_3}(t) &= P_4^{SS_2}(t)P_1^{EMS}(t) + P_4^{SS_2}(t)P_2^{EMS}(t),\\
P_5^{SS_3}(t) &= P_5^{SS_2}(t)P_1^{EMS}(t) + P_5^{SS_2}(t)P_2^{EMS}(t) + P_3^{EMS}(t).
\end{aligned}
$$

- L_Z-transforms for the entire DED system

$$
\begin{aligned}
L_z\{G^{DED}(t)\} &= \Omega_{f_{ser}}\left(L_z\{G^{SS_3}(t)\}, L_z\{G^{FPP}(t)\}\right) \\
&= \Omega_{f_{ser}}\left(P_1^{SS_3}(t)z^{5,500} + P_2^{SS_3}(t)z^{4,125} + P_3^{SS_3}(t)z^{2,750}\right. \\
&\quad \left. + P_4^{SS_3}(t)z^{1,375} + P_5^{SS_3}(t)z^0, p_1^{FPP}(t)z^{5,500} + p_2^{FPP}(t)z^0\right) \\
&= P_1^{DED}(t)z^{5,500} + P_2^{DED}(t)z^{4,125} + P_3^{DED}(t)z^{2,750} \\
&\quad + P_4^{DED}(t)z^{1,375} + P_5^{DED}(t)z^0.
\end{aligned}
\tag{2.27}
$$

where

$$
\begin{aligned}
P_1^{DED}(t) &= P_1^{SS_3}(t)p_1^{FPP}(t), \\
P_2^{DED}(t) &= P_2^{SS_3}(t)p_1^{FPP}(t), \\
P_3^{DED}(t) &= P_3^{SS_3}(t)p_1^{FPP}(t), \\
P_4^{DED}(t) &= P_4^{SS_3}(t)p_1^{FPP}(t), \\
P_5^{DED}(t) &= P_5^{SS_3}(t)p_1^{FPP}(t) + p_2^{FPP}(t).
\end{aligned}
$$

2.3.4 Multi-state Model for Multi-power Source Traction Drive in Norilsk-Type Ships

As one can see in Fig. 2.4, the multi-state model for Multi-Power Source Traction Drive in Norilsk-type ships may be presented as connected in series sub-system SS_3 with a gear box and a variable pitch propeller. Sub-system SS_3 consists of connected in parallel two sub-systems. Each sub-system consists of connected in series a Medium-Speed Diesel Engine, a fluid coupling and a clutch. Using the recursive derivation approach [8], we will present the whole system using the *Lz*-transform as follows:

$$
\begin{aligned}
L_z\{G^{SS_{11}}(t)\} &= \Omega_{f_{ser}}\left(L_z\{G^{MSDE_1}(t)\}, L_z\{G^{FC_1}(t)\}\right), \\
L_z\{G^{SS_{12}}(t)\} &= \Omega_{f_{ser}}\left(L_z\{G^{SS_{11}}(t)\}, L_z\{G^{CL_1}(t)\}\right) \\
L_z\{G^{SS_{21}}(t)\} &= \Omega_{f_{ser}}\left(L_z\{G^{MSDE_2}(t)\}, L_z\{G^{FC_2}(t)\}\right), \\
L_z\{G^{SS_{22}}(t)\} &= \Omega_{f_{ser}}\left(L_z\{G^{SS_{21}}(t)\}, L_z\{G^{CL_2}(t)\}\right) \\
L_z\{G^{SS_3}(t)\} &= \Omega_{f_{par}}\left(L_z\{G^{SS_{12}}(t)\}, L_z\{G^{SS_{22}}(t)\}\right) \\
L_z\{G^{SS_4}(t)\} &= \Omega_{f_{ser}}\left(L_z\{G^{SS_3}(t)\}, L_z\{G^{GB}(t)\}\right) \\
L_z\{G^{DGD}(t)\} &= \Omega_{f_{ser}}\left(L_z\{G^{SS_4}(t)\}, L_z\{G^{VPP}(t)\}\right)
\end{aligned}
\tag{2.28}
$$

Using the composition operators Ω_{fser} and Ω_{fpar} for sub-systems and elements, we obtain the following *Lz*-transforms:

- *Lz*-transforms for SS_{i1} subsystem i = 1,2

$$\begin{aligned} L_z\{G^{SS_{i1}}(t)\} &= \Omega_{f_{ser}}\left(L_z\{G^{MSDE_i}(t)\}, L_z\{G^{FC_i}(t)\}\right) \\ &= \Omega_{f_{ser}}\left(p_1^{MSDE_i}(t)z^{7,720} + p_2^{MSDE_i}(t)z^0, p_1^{MSb}(t)z^{7,720} + p_2^{MS}(t)z^0\right) \\ &= P_1^{SS_{i1}}(t)z^{7,720} + P_2^{SS_{i1}}(t)z^0. \end{aligned} \tag{2.29}$$

where

$$\begin{aligned} P_1^{SS_{i1}}(t) &= p_1^{MSDE_i}(t)p_1^{FC_i}(t), \\ P_2^{SS_1}(t) &= p_2^{MSDE_i}(t)p_1^{FC_i}(t) + p_2^{FC_i}(t). \end{aligned}$$

- *Lz*-transforms for SS_{i2} subsystem, i = 1, 2

$$\begin{aligned} L_z\{G^{SS_{i2}}(t)\} &= \Omega_{f_{ser}}\left(L_z\{G^{SS_{i1}}(t)\}, L_z\{G^{CL_i}(t)\}\right) \\ &= \Omega_{f_{ser}}\left(p_1^{SS_{i1}}(t)z^{7,720} + p_2^{SS_{i1}}(t)z^0, p_1^{CL_i}(t)z^{7,720} + p_2^{CL_i}(t)z^0\right) \\ &= P_1^{SS_{i2}}(t)z^{7,720} + P_2^{SS_{i2}}(t)z^0. \end{aligned} \tag{2.30}$$

where

$$\begin{aligned} P_1^{SS_{i2}}(t) &= P_1^{SS_{i1}}(t)p_1^{CL_i}(t), \\ P_5^{SS_2}(t) &= P_2^{SS_{i1}}(t)p_1^{CL_i}(t) + p_2^{CL_i}(t). \end{aligned}$$

- *Lz*-transforms for the SS_3 subsystem

$$\begin{aligned} L_z\{G^{SS_3}(t)\} &= \Omega_{f_{par}}\left(L_z\{G^{SS_{12}}(t)\}, L_z\{G^{SS_{22}}(t)\}\right) \\ &= \Omega_{f_{par}}\left(P_1^{SS_{12}}(t)z^{7,720} + P_2^{SS_{12}}(t)z^0, P_1^{SS_{22}}(t)z^{7,720} + P_2^{SS_{22}}(t)z^0\right). \end{aligned} \tag{2.31}$$

Using simple algebraic calculations of the powers of z as the sum of the values of the powers of corresponding terms, the whole system's L_Z-transform expression is as follows:

$$L_z\{G^{SS_3}(t)\} = P_1^{SS_3}(t)z^{15{,}440} + P_2^{SS_3}(t)z^{7{,}720} + P_3^{SS_3}(t)z^0 \tag{2.32}$$

where

$$P_1^{SS_3}(t) = P_1^{SS_{12}}(t)P_1^{SS_{22}}(t),$$
$$P_2^{SS_3}(t) = P_1^{SS_{12}}(t)P_2^{SS_{22}}(t) + P_2^{SS_{12}}(t)P_1^{SS_{22}}(t),$$
$$P_3^{SS_3}(t) = P_2^{SS_{12}}(t)P_2^{SS_{22}}(t)$$

- L_Z-transforms for the SS_4 subsystem

$$\begin{aligned} L_z\{G^{SS_4}(t)\} &= \Omega_{f_{ser}}\left(L_z\{G^{SS_3}(t)\}, L_z\{G^{GB}(t)\}\right) \\ &= \Omega_{f_{ser}}\left(P_1^{SS_3}(t)z^{15{,}440} + P_2^{SS_3}(t)z^{7{,}720}\right. \\ &\quad \left. + P_3^{SS_3}(t)z^0, p_1^{GB}(t)z^{15{,}440} + p_2^{GB}(t)z^0\right) \\ &= P_1^{SS_4}(t)z^{15{,}440} + P_2^{SS_4}(t)z^{7{,}720} + P_3^{SS_4}(t)z^0. \end{aligned} \tag{2.33}$$

where

$$P_1^{SS_4}(t) = P_1^{SS_3}(t)p_1^{GB}(t),$$
$$P_2^{SS_4}(t) = P_2^{SS_3}(t)p_1^{GB}(t),$$
$$P_3^{SS_4}(t) = P_3^{SS_3}(t)p_1^{GB}(t) + p_2^{GB}(t)$$

- L_Z-transforms for the DGD system

$$\begin{aligned} L_z\{G^{DGD}(t)\} &= \Omega_{f_{ser}}\left(L_z\{G^{SS_4}(t)\}, L_z\{G^{VPP}(t)\}\right) \\ &= \Omega_{f_{ser}}\left(P_1^{SS_4}(t)z^{15{,}440} + P_2^{SS_4}(t)z^{7{,}720} + P_3^{SS_4}(t)z^0,\right. \\ &\quad \left. p_1^{VPP}(t)z^{15{,}440} + p_2^{VPP}(t)z^0\right) \\ &= P_1^{DGD}(t)z^{15{,}440} + P_2^{DGD}(t)z^{7{,}720} + P_3^{DGD}(t)z^0. \end{aligned} \tag{2.34}$$

where

$$P_1^{DGD}(t) = P_1^{SS_4}(t)p_1^{VPP}(t),$$
$$P_2^{DGD}(t) = P_2^{SS_4}(t)p_1^{VPP}(t),$$
$$P_3^{DGD}(t) = P_3^{SS_4}(t)p_1^{VPP}(t) + p_2^{VPP}(t)$$

2.3.5 *Calculation of the Reliability Indices of Multi-power Source Traction Drives*

Using expression (2.3), the instantaneous availability for constant demand level w may be presented as follows:

- Winter period—75% demand level

$$\begin{aligned} A_{w\geq 4{,}125\mathrm{kW}}^{DED}(t) &= \sum_{g_i^{DED}\geq 4{,}125} P_i^{DED}(t) = P_1^{DED}(t) + P_2^{DED}(t), \\ A_{w\geq 11{,}580\mathrm{kW}}^{DGD}(t) &= \sum_{g_i^{DGD}\geq 11{,}580} P_i^{DGD}(t) = P_1^{DGD}(t). \end{aligned} \tag{2.35}$$

- Summer period—50% demand level

$$\begin{aligned} A_{w\geq 2{,}750\mathrm{kW}}^{DED}(t) &= \sum_{g_i^{DED}\geq 2{,}750} P_i^{DED}(t) = P_1^{DED}(t) + P_2^{DED}(t) + P_3^{DED}(t), \\ A_{w\geq 7{,}720\mathrm{kW}}^{DGD}(t) &= \sum_{g_i^{DGD}\geq 7{,}720} P_i^{DGD}(t) = P_1^{DGD}(t) + P_2^{DGD}(t). \end{aligned} \tag{2.36}$$

The instantaneous power performance (2.4) for Multi-Power Source Traction Drives can be obtained in the following manner:

$$\begin{aligned} E^{DED}(t) &= \sum_{g_i^{DED}>0} g_i^{DED} P_i^{DED}(t) = \sum_{i=1}^{4} g_i^{DED} P_i^{DED}(t) \\ &= 5,500 \cdot P_1^{DED}(t) + 4,125 \cdot P_2^{DED}(t) + 2,750 \cdot P_3^{DED}(t) + 1,375 \cdot P_4^{DED}(t), \\ E^{DGD}(t) &= \sum_{g_i^{DGD}>0} g_i^{DGD} P_i^{DGD}(t) = \sum_{i=1}^{2} g_i^{DGD} P_i^{DGD}(t) \\ &= 15,440 \cdot P_1^{DGD}(t) + 7,720 \cdot P_2^{DGD}(t). \end{aligned} \tag{2.37}$$

The instantaneous relative power performance for Multi-Power Source Traction Drives can be obtained in the following manner:

$$
\begin{aligned}
E_{\text{Rel}}^{DED}(t) &= \frac{1}{5,500} \sum_{g_i^{DED}>0} g_i^{DED} P_i^{DED}(t) = \frac{1}{5,500} \sum_{i=1}^{4} g_i^{DED} P_i^{DED}(t), \\
E_{\text{Rel}}^{DGD}(t) &= \frac{1}{15,440} \sum_{g_i^{DGD}>0} g_i^{DGD} P_i^{DGD}(t) = \frac{1}{15,440} \sum_{i=1}^{2} g_i^{DGD} P_i^{DGD}(t).
\end{aligned} \tag{2.38}
$$

The instantaneous power performance deficiency (2.5) of Multi-Power Source Traction Drives for the winter period can be obtained in the following manner:

$$
\begin{aligned}
D^{DED}(t) &= \sum_{i=1}^{5} P_i^{DED}(t) \cdot \max(4,125 - g_i, 0) \\
&= 1,375 \cdot P_3^{DED}(t) + 2,750 \cdot P_4^{DED}(t) + 4,125 \cdot P_5^{DED}(t), \\
D^{DGD}(t) &= \sum_{i=1}^{3} P_i^{DGD}(t) \cdot \max(11,580 - g_i, 0) \\
&= 3,860 \cdot P_2^{DGD}(t) + 11,580 \cdot P_3^{DGD}(t).
\end{aligned} \tag{2.39}
$$

The instantaneous power deficiency of Multi-Power Source Traction Drives for the summer period can be obtained in the following manner:

$$
\begin{aligned}
D^{DED}(t) &= \sum_{i=1}^{5} P_i^{DED}(t) \cdot \max(2,750 - g_i, 0) = 1,375 \cdot P_4^{DED}(t) + 2,750 \cdot P_5^{DED}(t), \\
D^{DGD}(t) &= \sum_{i=1}^{3} P_i^{DGD}(t) \cdot \max(7,720 - g_i, 0) = 7,720 \cdot P_3^{DGD}(t).
\end{aligned} \tag{2.40}
$$

The failure and repair rates (per year^{-1}) for each system's elements are presented in Tables 2.1 and 2.2.

Table 2.1 Failure and repair rates of elements in Amguema-type ships (per year^{-1})

	Failure rates	Repair rates
Diesel engine	2.2	73
Generator	0.15	175.2
Main switchboard	0.05	584
Electric energy converter	0.2	673
Electric motor	0.26	116.8
Fixed pitch propeller	0.01	125

Table 2.2 Failure and repair rates of elements in Norilsk-type ships (per year^{-1})

	Failure rates	Repair rates
Diesel engine	1.5	230
Gear box	0.11	195
Coupling	0.15	398
Clutch	0.11	467
Variable pitch propeller	0.11	92

Table 2.3 Reliability indices of multi-power source traction drives after a year of operations

	Amguema-type ships	Norilsk-type ships
Instantaneous availability of traction drive for winter demand	0.9899	0.9228
Instantaneous availability of traction drive for summer demand	0.9994	0.9969
Power performance	5321 kW	14,820 kW
Relative power performance	0.9674	0.9598
Power performance deficiency for winter demand	15.3 kW	322.0 kW
Power performance deficiency for summer demand	1.4 kW	24.2 kW

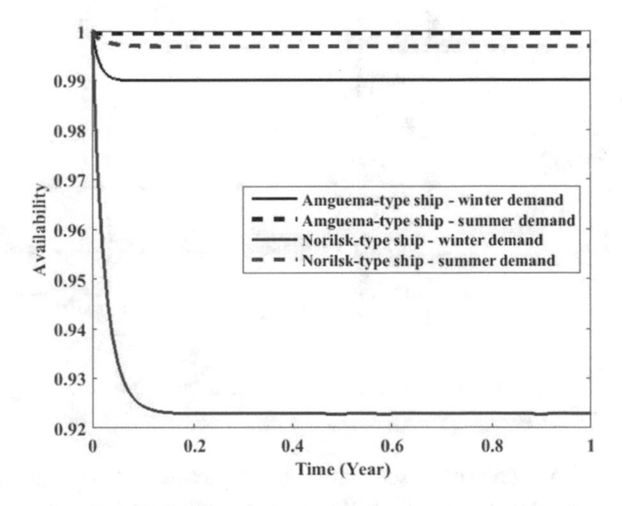

Fig. 2.6 Availability of traction drives in Amguema- and Norilsk-type ships for different constant demand levels

The calculated reliability indices of Multi-Power Source Traction Drives in Amguema- and Norilsk-type ships respectively after a year of operations are presented in Table 2.3 and Figs. 2.6, 2.7, 2.8, 2.9, 2.10.

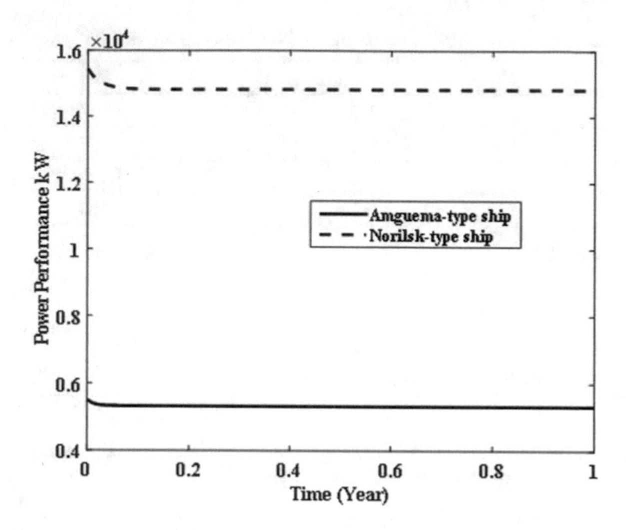

Fig. 2.7 Power performance of traction drives in Amguema- and Norilsk-type ships

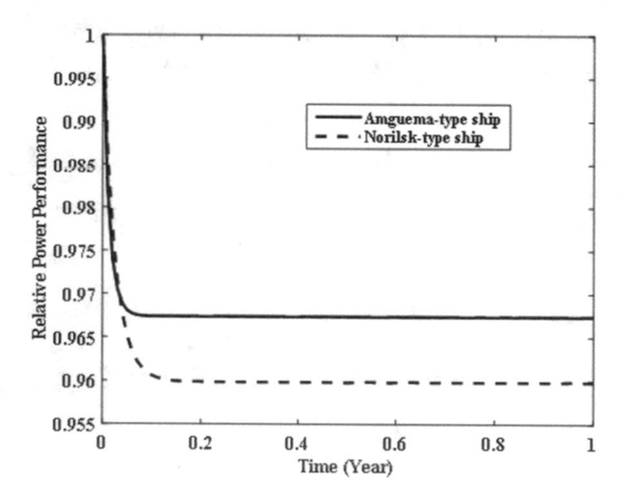

Fig. 2.8 Relative power performance of traction drives in Amguema- and Norilsk-type ships

As one can see, the instantaneous availability of the traction drives in Amguema-type ships during the winter and summer periods is greater than the instantaneous availability of the traction drive in Norilsk-type ships during the winter and summer periods.

In addition, the relative power performance of the traction drive in Amguema-type ships is greater than the relative power performance of the traction drive in Norilsk-type ships.

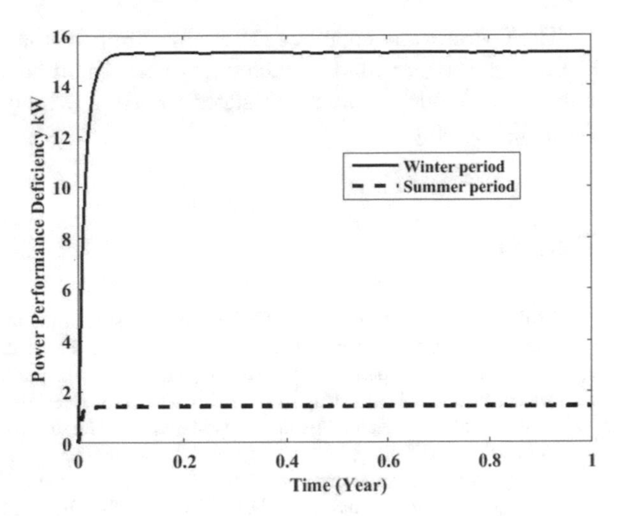

Fig. 2.9 Power performance deficiency of traction drive in Amguema-type ships

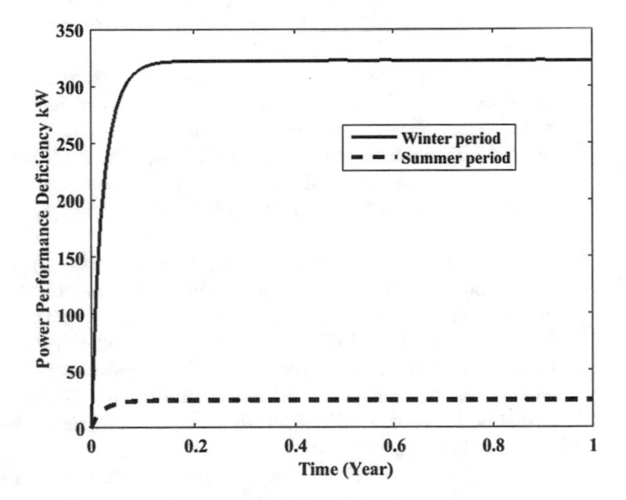

Fig. 2.10 Power performance deficiency of traction drive in Norilsk-type ships

2.4 Conclusion

In this Chapter, the L_Z-transform method was used for the evaluation of important parameters of the vehicle's operational sustainability—the availability, power performance and power performance deficiency of the multi-state Multi-Power Source Traction Drives in the two types of icebreaker ships considered here—Amguema and Norilsk. The results of calculations based on L_Z-transform correlate with the statistical data on the operational availability of diesel-electric and diesel-geared propulsion systems, as discussed in Chap. 1.

The L_Z-transform approach extremely simplifies the solution, in comparison with the straightforward Markov method, which would have required construction and solution of models with 8,192 states for Amguema-type ship and 256 states for Norilsk-type ship.

References

1. Bolvashenkov I, Herzog HG (2016) Use of stochastic models for operational efficiency analysis of multi power source traction drives. In: Proceedings of the second international symposium on stochastic models in reliability engineering, life science and operations management, (SMRLO16), 15–18 February 2016, Beer Sheva, Israel, pp 124–130
2. Bolvashenkov I, Kammermann J, Willerich S, Herzog HG (2016) Comparative study of reliability and fault tolerance of multi-phase permanent magnet synchronous motors for safety-critical drive trains. In: Proceedings of the international conference on renewable energies and power quality (ICREPQ16), 4–6 May 2016, Madrid, Spain, pp 1–6
3. Frenkel I, Bolvashenkov I, Herzog HG, Khvatskin L (2016) Performance availability assessment of combined multi power source traction drive considering real operational conditions. Transp Telecommun 17(3):179–191
4. Frenkel I, Bolvashenkov I, Herzog HG, Khvatskin L (2017) Operational sustainability assessment of multi power source traction drive. In: Ram M, Davim JP (eds) Mathematics applied to engineering. Elsevier, London, pp 191–203
5. Frenkel I, Bolvashenkov I, Herzog HG, Khvatskin L (2017) Lz-transform approach for fault tolerance assessment of various traction drives topologies of hybrid-electric helicopter. Recent Advances in multistate system reliability: theory and applications. Springer, London, pp 321–362
6. Jia H, Jin W, Ding Y, Song Y, Yu D (2017) Multi-state time-varying reliability evaluation of smart grid with flexible demand resources utilizing Lz transform. In: Proceedings of the international conference on energy engineering and environmental protection (EEEP2016), IOP Publishing, IOP Conference Series: Earth and Environmental Science, vol 52
7. Lisnianski A (2012) Lz-Transform for a discrete-state continuous-time markov process and its applications to multi-state systems reliability. In: Lisnianski A, Frenkel I (eds). Recent advances in system reliability: signatures, multi-state systems and statistical inference, Springer, London, pp 79–97
8. Lisnianski A, Frenkel I, Ding Y (2010) Multi-state system reliability analysis and optimization for engineers and industrial managers. Springer, London
9. Ushakov I (1986) A universal generating function. Sov J Comput Syst Sci 24:37–49
10. Yu H, Yang J, Mo H (2014) Reliability analysis of repairable multi-state system with common bus performance sharing. Reliab Eng Syst Safety 132:90–96
11. Natvig B (2011) Multistate systems reliability, theory with applications. Wiley, New York
12. Trivedi K (2002) Probability and statistics with reliability, queuing and computer science applications. Wiley, New York

Chapter 3
The Two-Step Approach to the Selection of a Traction Motor for Electric Vehicles

Abstract This chapter presents the two-step comparative analysis of electrical traction machines. The stable permanent demand of electrical drives requires appropriate selection of electrical machines to meet specific requirements. Different vehicles, such as aircraft, cars, ships and trains are considered here. Each kind of electric vehicle needs an electrical traction machine with unique parameters. The results show that the selection of the appropriate electrical machine will depend on the type of vehicle. Furthermore, in most cases there is always more than one appropriate machine type. Thus, the type of electrical traction machine has to be defined for each vehicle and a comparative analysis is a crucial tool here.

Keywords Traction motors · Electrical propulsion system · Lifecycle costs analysis · Operational damage · Electric vehicle · Aircraft · Ship
Train

3.1 Introduction

Currently, there are many different types of electric vehicles that have been designed and constructed [7, 8, 19]. Different electrical machine types must be investigated so that one can determine what type will be appropriate for the specific vehicle under consideration (car, aircraft, ship or train). Since the requirements will differ from one vehicle to the next, the present chapter focuses on a comparison of different machine types for each type of vehicle.

First, the objects of comparison, vehicles (hybrid and electric) and traction electric motors (TEMs) are explained. We then discuss the ultimate results of the process of selection of the most appropriate motor for the vehicle being considered and draw conclusions as to the usability of the methodology.

The two-step selection of the optimal type of traction electric motor consists of preliminary and final levels of evaluation. First, a rough estimate is made and the two best options are selected. In the final level of evaluation, the most suitable motor is selected.

I. Bolvashenkov et al., *Safety-Critical Electrical Drives*, SpringerBriefs in Electrical and Computer Engineering, https://doi.org/10.1007/978-3-319-89969-5_3

3.2 Electric Vehicles

The electrification of different types of vehicles is widely discussed today because of the many benefits involved, such as:

- reduced emissions;
- increased operational efficiency;
- reduced energy consumption;
- increased reliability and fault tolerance;
- simplified control;
- reduced airport noise;
- simplified autonomous implementation.

What type of traction electric motor is best suited for the electric propulsion system of a vehicle? Unfortunately, the best, optimal solution for all types of transport does not exist yet. Therefore, for each type of transport vehicle, this task must be solved individually. The optimal type of traction electric motor will depend on the specific vehicle type, on their operating conditions and parameters and on their unique requirements and limitations. A large number of different parameters should be considered for solving the multifactor optimization problem. This can be done only through a systematic approach. The features of certain types of electrified vehicles are briefly considered below.

3.2.1 *Aircraft*

This category of vehicle includes airplanes (international carriers, regional planes, special purpose aircraft, piloted and pilotless aircraft, etc.) as well as helicopters (from heavy to light types, piloted and pilotless, etc.). At the same time, if we are focusing on airplanes, step-by-step electrification must be envisaged (from the hybrid version to the fully electric version); for helicopters, due to their limited space, the full electrical version must be considered.

The specific operational modes of the helicopter, in contrast with the airplane, require more power and a higher energy density in the traction drive, and there is thus a need to install a more powerful electric traction motor with a larger capacity electric energy source or energy storage. We must also take into consideration the maximum takeoff weight (MTOW) of the helicopter and the space for the installation of the electric traction drive components. Thus, it is necessary to consider the following features of helicopter operation:

- low autonomy and, accordingly, short duration flights;
- lack of structural redundancy;
- inability to repair during flight;
- the most stringent restrictions on weight and fault tolerance.

Thus, the correct choice of the optimal type of helicopter traction electric motor is a highly important task, the solution of which will be considered in the following sections.

3.2.2 Cars

During our investigations, the following features of the operation of passenger hybrid/electric cars were taken into account:

- small average annual mileage of the car;
- low value of average utilization rate;
- lack/partial availability of structural redundancy;
- relatively rigid restrictions on weight and dimensions of the traction drive;
- possibility of correcting small malfunctions during operations;
- high maintainability and accordingly a relatively high availability factor.

It should be noted that in this chapter, in the category of cars, the features of truck and bus operations were not considered.

3.2.3 Ships

Given the specific operational conditions of ships, the following features were taken into account for the design of their electric propulsion systems:

- high autonomy of the ship's operations;
- availability of structural and functional redundancy;
- high value of the rate of technical use;
- high maintainability of electric propulsion system;
- possibility of repairs during operations;
- high survivability of the propulsion system;
- absence of tight restrictions on the weight and dimensions of the traction drive.

3.2.4 Trains

In studying the specific characteristics of this mode of vehicle, the following features of its operations were taken into account:

- limited autonomy of operations;
- availability of structural redundancy;
- high coefficient of technical use and running time;

- relatively high maintainability of the propulsion system;
- possibility of minor repairs during operations;
- absence of tight restrictions on the weight and dimensions of the traction drive.

3.3 Traction Electric Motors

This section provides an overview of the types of machines we compared. The different electrical machines are briefly described in terms of properties, special features, and technical data.

For the comparative analysis, the following machines were taken into account: the induction machine (IM), the switched reluctance machine (SRM) and the synchronous machine (SM), including the permanent magnet SM with distributed windings (PSM-d) and the permanent magnet SM with concentrated windings (PSM-c). Figure 3.1 presents the possible realizations for each machine type [28].

3.3.1 Induction Machine

The cross section of an induction machine is shown in Fig. 3.1a. The main advantage of induction machines is high availability and relatively low maintenance cost [28]. Induction machines produce a low level of noise or vibrations compared to other machine types, the production cost is relatively low and the ability to operate in hostile environments is high. However, the power operation is limited, i.e., efficiency decreases at lower speeds and the power decreases at higher speeds. In addition, the induction motor is almost impossible to partition, i.e., it is almost impossible to divide one powerful machine into several low-power ones, because it is impossible to make unconnected stator phases that will be controlled by a separate electric converter completely independent of the others. Therefore, to ensure

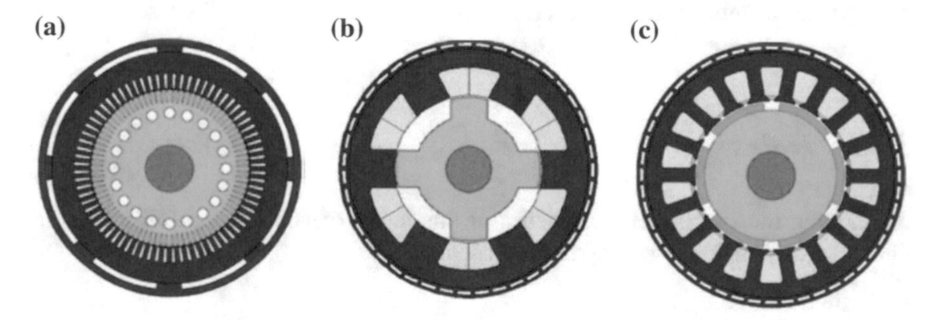

Fig. 3.1 Cross sections of different electrical machines: **a** induction machine, **b** switched reluctance machine, **c** synchronous machine with permanent magnets

the high power of the induction motor's converter, it is necessary to increase its voltage, that is, to use a high-voltage converter. In the case of sectioned electric drives, it is advisable to take a low-voltage converter with a relatively small nominal current and supply the necessary amount of submodules; in the case of traction drives with an induction motor, there should be only one electric converter.

3.3.2 Switched Reluctance Machine

The cross section of the switched reluctance machine (SRM) is presented in Fig. 3.1b. It includes the typical rotor without excitation coils or permanent magnets (PM) in the rotor. Some of the advantages of the SRM are low maintenance and low production cost, as there is no excitation needed in the rotor. Consequently, rotor losses are minimized. Moreover, the control is simple and the reliability is high compared to other machines. In addition, the SRM provides thermal stability and has a relatively high overload capability.

Due to the simplicity of the design, this kind of motor turns out to be cheaper than the classic induction motor. In addition, the SRM is easy to make in multi-phase and multi-section versions, where we divide the control of one motor into several independent converter submodules working in parallel. We can thus increase the reliability of the motor drive; thus, for example, the failure of one electric converter will not lead to the total failure of the whole electric drive, and the remaining submodules will work for a while with a slight overload.

However, the SRM has high torque ripples and therefore an increased generation of noise and vibrations. Due to the specific design of the motor and the impulse nature of the current, it is difficult to obtain a stable torque, and as a rule, the torque in the SRM pulsates. That feature limits the applicability of this motor type for vehicular traction drives. In addition, the pulsating nature of the traction torque has a negative effect on the lifetime of the motor bearings. This problem can be partially solved through a special profiling of the shape of the phase current, as well as through an increase in the number of phases. As discussed in [22], these machines are limited if we need torque at low speeds, because of the factor of saturation.

3.3.3 Permanent Magnet Synchronous Machine

The cross section of the synchronous machine (SM) is shown in Fig. 3.1c. It is realized as a PSM with a surface PM. Permanent magnets provide the necessary magnetization of the rotor without corresponding losses, and that feature increases the efficiency of this type of motor in comparison with inductive ones. Since expensive rare-earth metals are required for the production of magnets, until now the price of such engines has been quite high and demand has been higher than supply. Nevertheless, in recent years, with the increase in PSM production, there has been a significant decline in prices.

Compared to all the other machine types, the PSM has high power density and a high level of efficiency around the nominal speed. Additionally, due to the missing rotor windings, there are no copper losses; thus, less cooling is required [22, 28]. A PSM can be implemented in a smaller volume for the same speed and power than an SRM [22]. Due to the use of rare earths, the price is high and this is an important factor considering the PSM, even though it is not very decisive for some applications. Distributed and concentrated windings in the PSM have been taken into account for the comparison.

The PSM is optimal for cases where a large control range is not required. For example, such a kind of motor is ideal for drives with a fan load characteristic. In that version of the PSM application, the rotation speed changes are relatively small, with a maximum of two speed changes, because the airflow is weakened in proportion to the square of the speed.

The disadvantages of this type of motor include the risk of demagnetization at high temperatures, a situation that, however, is rarely encountered in practice. In addition, the problems associated with motor repairs must be taken into account. There are two primary problems: There are strong magnets in the rotor and the process of extracting the rotor from the stator is complex and requires the use of special tools.

The abovementioned characteristics of electrical machines complicate the selection of a suitable machine, due to the high complexity and interdependence of parameters and the mixed advantages and disadvantages. The following sections introduce the parameters for a preliminary comparative analysis.

3.4 First Step—A Preliminary Comparative Analysis

Ten parameters, used in the comparison, will be mentioned and briefly explained in this section. Each parameter has a certain influence on a system's behavior and hence has to be considered. Furthermore, the priority of parameters can differ from one application to another. Therefore, parameters have to be weighted in order to meet the different requirements for different applications.

3.4.1 Parameter Identification

The main goal of the applications is to transport persons or cargo, and the transport has to provide safety, usability, and efficient operations. Therefore, the indicators for comparative evaluation must take into account the requirements and conditions for the future operation of the applications. Additionally, the complexity, cost, and maintainability of the system play an important role for the manufacturer.

Furthermore, the parameters need to be independent, not correlating with each other, since a correlation of parameters will shift the results. Only the lower limits of values as determined by considerations and calculations and as noted in existing publications were taken into account for evaluation.

Thus, the pre-design is supposed to meet the requirements even though at an early stage of a design process some effects might not be considered yet. In some cases, adequate assumptions have to be made. The abovementioned procedure leads to the following parameters:

- efficiency,
- power density,
- Mean Time to Failure (MTTF),
- Mean Time to Repair (MTTR),
- reparability,
- fault tolerance,
- noise,
- volume,
- complexity,
- cost.

3.4.2 Determination of Parameters

In this section, the determination of parameters will be shown in order to provide an example for the evaluation of reliability and fault tolerance.

3.4.2.1 Reliability

For aircraft applications, the reliability and fault tolerance of the main electric motors are extremely important characteristics of electrical machines and are often decisive considerations. In other words, reliability and safety define the design features and the structural connections of electrical drive trains.

Therefore, the most accurate comparative analysis and evaluation of reliability indices at the design stage of various electrical machines, which are a part of the electrical drive train and which take into account future operating conditions, can be considered an important problem. The main faults in electric machines can be classified as in [27]:

- stator winding faults;
- broken rotor bar or end-ring faults on the IM;
- static/dynamic air gap irregularities (rotor eccentricity);
- bearing failures;
- defects of the permanent magnets of the PSM.

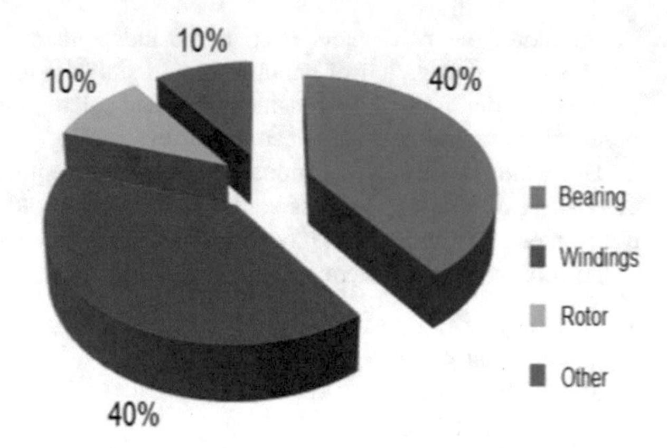

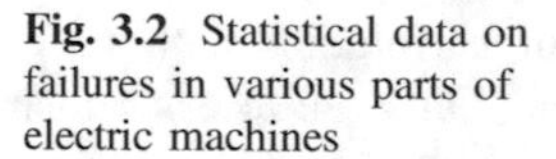
Fig. 3.2 Statistical data on failures in various parts of electric machines

Each fault disrupts the motor's normal operations, producing several symptoms, such as unbalanced line currents and air gap voltages, torque and speed pulsations, decreased efficiency and lower torque, excessive heating, and increased losses. The operating experience of electrical machines indicates that the most vulnerable elements of electrical machines are the stator windings and the bearings [15, 20, 21]. Statistical data on failures in various parts of electric machines is shown in Fig. 3.2.

Considering the above data charts, the failure rate of electric machines λ_{EM} can generally be determined by the expression:

$$\lambda_{EM}(t) = \lambda_S(t) + \lambda_R(t) + \lambda_B(t). \tag{3.1}$$

where λ_S, λ_R, and λ_B are the failure rates of parts of the electrical machine (stator, rotor and bearing) respectively. The probability of failure-free operations of the system is defined as:

$$P_{EM}(t) = e^{-(\lambda_{SO} + \lambda_{RO} + \lambda_{BO})t} \tag{3.2}$$

where λ_{SO}, λ_{RO}, and λ_{BO} are the failure rates of the main parts of electrical machines, which are calculated considering defined operating conditions with correction factors k_i, $i = 1, \ldots m$ for each operational mode:

$$\lambda_{SO}(t) = \lambda_{Sm} \sum_{i=1}^{m} k_i t_i / t \tag{3.3}$$

$$\lambda_{RO}(t) = \lambda_{Rm} \sum_{i=1}^{m} k_i t_i / t \tag{3.4}$$

$$\lambda_{BO}(t) = \lambda_{Bm} \sum_{i=1}^{m} k_i t_i / t, \tag{3.5}$$

Table 3.1 Reliability values for various machine types

Motor type	*P* (100)	*P* (1,000)	*P* (10,000)	*P* (100,000)
IM	0.9998	0.9976	0.9751	0.8710
SRM	0.9999	0.9989	0.9782	0.8961
PSM	0.9999	0.9997	0.9811	0.9230

where λ_{Sm}, λ_{Rm}, and λ_{Bm} are the mean values of the failure rates of every part of electric machines in "ideal" laboratory conditions and t_i, $i = 1, \ldots, m$ are the durations of each mode of operation.

Based on statistical data, we estimated the reliability indices for the IM, SRM and PSM, with a given level of power (600 kW) and a given rotational speed (400 rpm). Table 3.1 summarizes the calculated values for various prediction intervals. Those results are presented graphically in Fig. 3.3.

3.4.2.2 Fault Tolerance

To evaluate the level of fault tolerance of the traction motor for an electrical helicopter, three characteristics of electric machines, which have a major impact on its functioning in fail operation mode, were analyzed:

- overload capacity;
- partial load operational mode;
- torque ripples in case of failure.

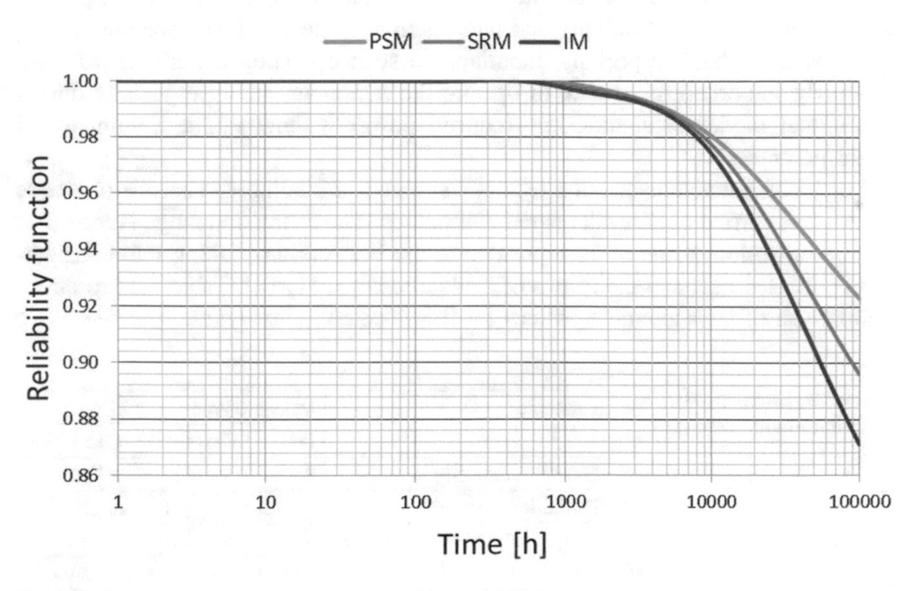

Fig. 3.3 Reliability functions of the IM, SRM and PSM

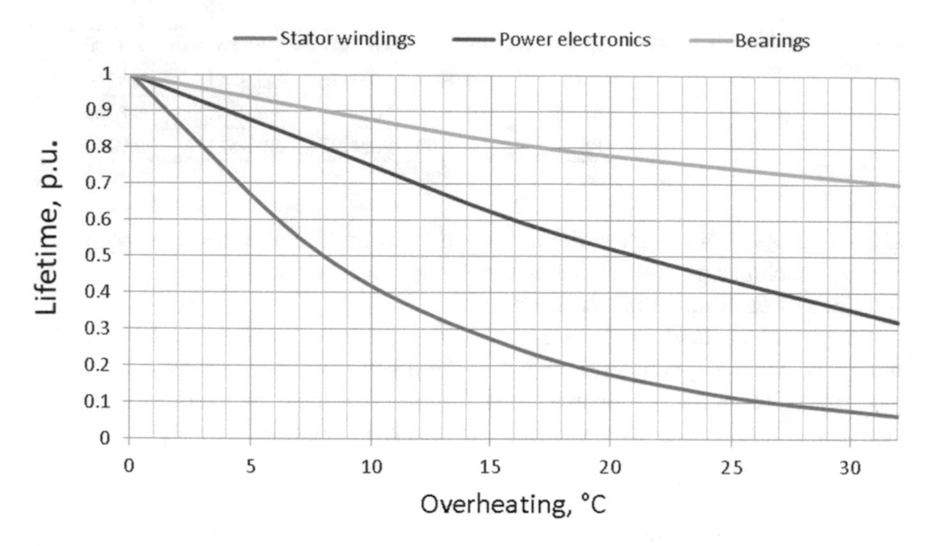

Fig. 3.4 Overheating influence on the lifetime of the components of an electric drive

During the normal (failure-free) operational mode, the electric motor can endure a short-term overload because its thermal capacity is sufficiently large. In failure cases, the situation changes dramatically.

The largest number of fail operational modes are caused by technological electric overloads. The consequences of an overload are the overcurrent and overheating of the electrical machine, which leads to a reduction of the reliability indices of the motor and a decrease of its lifetime, as can be seen in Fig. 3.4 [4].

For the traction motor of the electrical helicopter, considering the tight requirements of drive reliability and fault tolerance, the overload capability in fail operational modes is especially important. In such operating conditions it is also extremely important to be able to operate the helicopter stably especially during operational modes of significantly reduced power without undue asymmetry of motor parameters.

For a comparative evaluation of this parameter, a qualitative analysis of torque ripples was carried out for fail operational modes. The estimation range is from 1 to 10, the maximum value of 10 representing the best option. Taking into consideration collected data and the results of calculations [3, 11, 18], Table 3.2 presents a comparison of fault tolerance indices for the selected motor types.

Table 3.2 Comparative analysis of fault tolerance

Parameter	Machine type			
	IM	SRM	PSM-d	PSM-c
Overload capacity	8	8.5	9	10
Partial load mode	8	8	10	10
Torque ripple in failure case	8	7.5	10	9
Total/average value	24/8	24/8	29/9.7	29/9.7

3.4.2.3 Priority Coefficient

Weighting factors are needed in order to assure an appropriate evaluation of each application. Parameters are weighted roughly through priorities, which can be seen in Table 3.3. In addition to these pre-evaluated priorities, the priority coefficients K_p are determined and are presented in Tables 3.4, 3.5, 3.6 and 3.7.

The factors are derived from literature research, expert knowledge, and scientists' experience. As aforementioned, safety, usability and efficient operations are decisive factors regarding the electrical drive train. Therefore, parameter efficiency, MTTF, MTTR, reparability, and fault tolerance get the highest priority coefficient.

Table 3.3 Weighting factors for different vehicle types

Parameter	Vehicle type			
	Aircraft	Car	Ship	Train
Efficiency	++	++	++	++
Power density	++	+	+	+
MTTF	++	++	++	++
MTTR	x	++	++	++
Reparability	x	++	++	++
Fault tolerance	++	++	++	++
Noise	x	+	o	+
Volume	++	+	+	+
Complexity	o	+	+	+
Cost	+	++	++	++

x: not applicable

Table 3.4 Electrical machine types for aircraft application

Parameter	Machine type				
	K_p	IM	SRM	PSM-d	PSM-c
Efficiency	0.9	8	8	10	10
Power density	1.0	7	7.5	10	10
MTTF	1.0	10	10	10	10
MTTR	–	–	–	–	–
Reparability	–	–	–	–	–
Fault tolerance	1.0	8	7.5	9.5	10
Noise	–	–	–	–	–
Volume	0.9	7	8	10	9.5
Complexity	0.3	9	10	8	8
Cost	0.2	9	10	8	7.5
TOTAL/(without K_p)		58.0	61.0	65.5	65.0
Relative value		0.88	0.93	1.00	0.99
TOTAL/(with K_p)		42.3	43.6	50.6	50.4
Relative value		0.84	0.86	1.00	0.99

Table 3.5 Electrical machine types in automotive application

Parameter	Machine type				
	K_p	IM	SRM	PSM-d	PSM-c
Efficiency	1.0	8	8.5	10	10
Power density	0.8	7	7.5	10	10
MTTF	1.0	9	9.5	10	10
MTTR	1.0	10	10	9	9.5
Reparability	1.0	10	10	9	9
Fault tolerance	1.0	8	8	9.5	10
Noise	0.8	9	8	10	9.5
Volume	0.7	8.5	8	10	9.5
Complexity	0.8	9.5	10	8	8
Cost	1.0	10	10	8	7.5
TOTAL/(without K_p)		89.0	89.5	93.5	93.0
Relative value		0.95	0.96	1.00	0.99
TOTAL/(with K_p)		81.0	80.0	85.1	84.9
Relative value		0.95	0.96	1.00	0.99

Table 3.6 Electrical machine types in maritime application (ships)

Parameter	Machine type				
	K_p	IM	SRM	PSM-d	PSM-c
Efficiency	1.0	8	6	10	9.5
Power density	0.6	8	6	10	10
MTTF	1.0	9	9	10	10
MTTR	1.0	10	10	9.5	9.5
Reparability	1.0	10	10	9.5	9.5
Fault tolerance	1.0	8	8	9	10
Noise	0.5	9	8	10	9.5
Volume	0.6	8	7	10	9.5
Complexity	0.8	10	10	8	8.5
Cost	1.0	10	10	8	7.5
TOTAL/(without K_p)		90.0	84.0	94.0	93.5
Relative value		0.96	0.89	1.00	0.99
TOTAL/(with K_p)		77.1	72.8	79.6	78.9
Relative value		0.97	0.91	1.00	0.99

Table 3.7 Electrical machine types in railway application

Parameter	Machine type				
	K_p	IM	SRM	PSM-d	PSM-c
Efficiency	0.9	8	8	10	9.5
Power density	1.0	7	7.5	10	10
MTTF	1.0	9	9.5	10	10
MTTR	1.0	10	10	9.5	9.5
Reparability	1.0	10	10	9.5	9.5
Fault tolerance	1.0	8	8	9	10
Noise	1.0	9	8	10	10
Volume	0.8	8	8	10	9.5
Complexity	0.3	10	10	8	8
Cost	0.2	10	10	8	7.5
TOTAL/(without K_p)		89.0	89.0	94.0	93.5
Relative value		0.95	0.95	1.00	0.99
TOTAL/(with K_p)		42.3	43.6	50.6	50.4
Relative value		0.95	0.95	1.00	0.99

3.5 Results of the First Step for Various Vehicular Applications

In this section, examples are given of the applications of electric aircraft, electric cars, electric ships, and electric trains and a comparison is made of the different machine types for each application in terms of parameters and the determination of priority coefficients.

3.5.1 *Electric Aircrafts*

Electric aircraft means fully electrified aircraft with direct drive and no gear box. Since weight plays an important role and a gear box usually accounts for a high percentage of weight as far as the drive train is concerned, these machines are characterized as low speed and high torque machines.

Furthermore, reliability, fault tolerance, and power density have to be treated as highly important factors. That is the reason why the PSM is considered to be the best choice for aircraft, as can be seen in Table 3.4. Cost and complexity are not important in the current investigation stage. Up to now there have been feasibility studies for aircraft, and the provision of safety, efficiency and power density is considered to be increasingly important [2, 23, 26].

3.5.2 Electric Cars

The results presented in Table 3.5 indicate that, for the automotive application, i.e., electric cars, different types of electrical machines can be used. Considering the car companies around the world, many European, American and Japanese companies use a PSM for hybrid electric cars, but some companies prefer the IM or SRM [22]. Figure 3.5 shows the distribution of the use of electrical machines in hybrid cars. The distribution of the relative weighting factor, which is more or less equal, is presented in Table 3.5. This means that the choice of the appropriate machine can depend on other factors, such as production knowhow, political aspects, and so on [12].

3.5.3 Electric Ships

The traction electric motors of ships are characterized by high power (over 1,000 kW) and a low rotational speed of the propeller shaft. Stringent requirements regarding the fault tolerance of the main traction motor are imposed for certain (but not for all) operational modes: passage through straits, channels, and other narrow waterways, in areas with difficult navigation conditions and in harsh weather and ice conditions. However, considering the less rigid requirements for dimensional characteristics (volume and weight), compared with aircraft equipment, there is significant potential for implementing structural and functional redundancy. Taking into account the fact that fuel cost accounts for 70% of ship operating expenditures, the efficiency of the motor is a highly important consideration [6]. Hence, the most promising machine types for marine electric propulsion systems are the PSM (e.g., produced by ABB) and the IM (manufactured by Siemens), see Table 3.6.

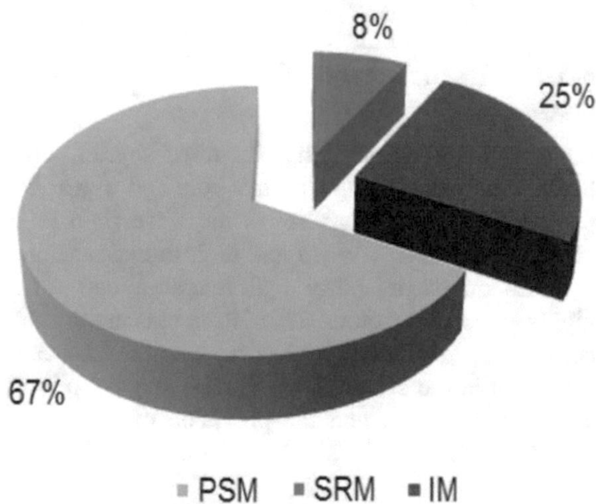

Fig. 3.5 Distribution of the use of electrical machines in hybrid electric cars

3.5.4 *Electric Trains*

Considering the modes of operation, design features and technical parameters of electric drive trains, the requirements for railway application lie between the automotive and the maritime applications. The power of the single traction electric motor is usually less than 1,000 kW and the rotational speed (without gears) is between 2,000 and 3,000 rpm. The requirements for weight, volume, size, and noise characteristics are more stringent than the requirements imposed on a ship's traction motors. At the same time, they are less stringent, in comparison with cars. Using a multi-aggregate scheme of the railway's electric propulsion, in which every locomotive with a motor (or motors) is a practically independent propulsion unit, we can minimize time and cost to implement an optimal maintenance strategy with the required level of redundancy and, accordingly, the required degree of fault tolerance.

The data presented in Table 3.7 shows that, despite the rather close total values of the complex index, the PSM has an advantage over the IM and the SRM. Nevertheless, today the PSM (Alstom) and the IM (Siemens) are used as traction motors. Also Japanese companies produce electric railway vehicles with the PSM and the IM [14].

Thus, the main challenge in selecting the most optimal type of traction motor has not yet been completely solved, and the developers often base themselves on their own experience in the design of electric motors.

For the second step in the comparative analysis, we propose a method of refined analysis, considered below, for the main traction motor of a ship's electric propulsion system.

3.6 Second Step—Comparative Analysis

The main disadvantage of existing methods for the choice of the optimal traction motor for electric vehicles is the use of technical and abstract parameters (such as the coefficient of technical level, technical excellence, etc.) or economic parameters (capital expenditures, specific cost, life cycle cost, etc.) as the universal criteria for comparison. Such parameters do not take into account specific operational conditions. Therefore, for an objective complex evaluation of the operational efficiency of the compared alternatives, it is necessary to carry out a statistical analysis of the operational modes, to create stochastic models of the functioning of the vehicles in real operational conditions and to simulate the compared variants in identical operating conditions.

Considering the complexity of the problem, its highly probabilistic nature and the presence of a number of uncertainties, the most objective decision can be found exclusively on the basis of a systematic approach. Cost-effective operations and the maintenance of a traction motor and the whole electric drive system require attention not only to individual units of the propulsion system, but to the system as a whole. A systems approach analyzes both the performance and demand sides of the system and how they interact, essentially shifting the focus from individual components to total system performance.

A common engineering approach is to break down a system into its basic components or units, to optimize the selection or design of those components, and then to assemble the system. One advantage of this approach is its simplicity. One disadvantage is that this approach ignores the interaction of the components. For example, a larger than necessary motor gives a safety factor and ensures that the motor can provide enough torque to meet the needs of the application. However, an oversized motor can also create performance problems with the driven equipment. In certain circumstances, an oversized motor can even compromise the reliability of both the components and the entire system.

In a component approach, an engineer employs a particular design that meets the unique requirements of a specific component. In a systematic approach, the entire system is evaluated in order to determine how end-use requirements can be provided mostly effectively and efficiently.

The study of different decision-making methods and of the comparative evaluation of complex technical systems, such as life cycle cost analysis (LCCA) [13], cost-benefit analysis (CBA) [1, 25], and multi-criteria analysis (MCA) [10, 29] for the purpose of choosing between alternatives, allow us to create a universal technique for the comprehensive assessment of traction electric motors [5, 9, 16, 17].

3.6.1 Brief Description of Methodology

3.6.1.1 Generalized Criteria

Chapter 1 provided a complete description of the method used in the comprehensive comparative assessment of the compliance of a vehicle's propulsion systems with planned operating conditions. According to [10] and the requirements of the systematic approach, the target function of electric vehicles is the "sustainable, effective timely delivery of the required amount of cargo or passengers." In accordance with the chosen target function, all the parameters are divided into two groups: those parameters which relate to the direct fulfillment of the target function of the vehicle—that is, to the criteria of "usefulness"—and those parameters which are related to the cost and damage involved in the fulfillment of the target function —that is, to the criterion of "payment for usefulness."

Thus, "usefulness" is a non-financial analog of the benefits criterion of CBA and "payment for usefulness" is a near analog of the financial criterion of LCCA.

Based on the proposed technique and taking into account the requirements and conditions of future operations, the vehicle's transportation productivity A was chosen as the criterion of "usefulness" for the vehicle's entire lifetime:

$$A = DVt_d. \tag{3.6}$$

$$t_d = t_o k_d. \tag{3.7}$$

where D is the amount of transported cargo or passengers, V is the average operational speed; t_o and t_d are respectively operational and driving time in hours, and k_d is the driving time rate.

As a second complex criterion C, "payment for usefulness," was accepted as the sum of capital and operating cost and damage for the entire lifetime of the vehicle:

$$C = C_{CAP} + C_{CONST} + C_F + C_{RAC} + C_{EAC}. \tag{3.8}$$

where C_{CAP} is capital cost, C_{CONST} the fixed operating cost (personnel, navigation fees, taxis, insurance, etc.), C_F the operating cost for fuel and oil, C_{RAC} the reliability-associated cost [16] and C_{EAC} the ecology-associated cost [5, 30].

The values of the parameters that are included in these two generalized criteria are dependent on the operational conditions of the vehicle in varying degrees. From the point of view of the operating conditions, the most informative parameters are the vehicle speed, the operating cost for fuel and oil, the reliability-associated cost and the ecology-associated cost.

3.6.1.2 Local Parameters

Considering the probabilistic nature of the vehicle movement process, it was advisable to introduce the value of the operational speed of the vehicle in the form of a multi-factor regression model (3.9).

$$V = f(x_1, x_2, \ldots, x_n). \tag{3.9}$$

where $x_1, x_2, \ldots, x_n$ are weakly correlated factors that affect the operational speed of the vehicle, such as the amount of transported cargo or passengers, the power of the main traction motor, the outside temperature, the strength and direction of the wind, etc.

It should be noted that the regression model takes into account factors explicitly included in the model, metric parameters of "usefulness," and factors not included explicitly, such as the type of vehicle design, propulsive quality, maneuverability and other unmeasurable factors.

Thus, for the calculation of the generalized criterion of "usefulness," it is necessary to determine the probability distribution (or average operational values) of these factors for the investigated operational area.

The calculation of values of C_{CAP} and C_{CONST} is a relatively simple problem and these values are not dependent on the vehicle's operating conditions. The values of C_F and C_{EAC} are calculated based on the Markov model of energy generation E of diesel engines for the entire period of the vehicle's operation:

$$C_F = c_f \sum_j e_j g_j. \tag{3.10}$$

where c_f is the specific fuel cost, e_j the energy generation in the j-mode and g_j the specific fuel consumption in the j-mode.

$$C_{EAC} = C_F \sum_i k_i d_i. \tag{3.11}$$

where k_i is the content of the harmful i-component in the burned fuel, C_F the amount of the burned fuel and d_i the specific damage from the combustion of the i-component of fuel.

For example, regarding trains and ships, for the further forecast of the value of C_{RAC}, the Markov model of reliability of the repairable system [3, 5, 16, 17] is effective, as can be seen in Fig. 3.6.

Using the Markov model, based on statistical operational data, we defined the availability coefficient of propulsion system K_A for the vehicle's lifetime. The value of the reliability-associated cost (RAC) was calculated using the formula:

$$C_{RAC} = c_{UR} t_o k_d (1 - K_A). \tag{3.12}$$

where c_{UR} is the average specific cost of ship/train downtime due to the propulsion system's unreliability and K_A the coefficient of the propulsion system's availability.

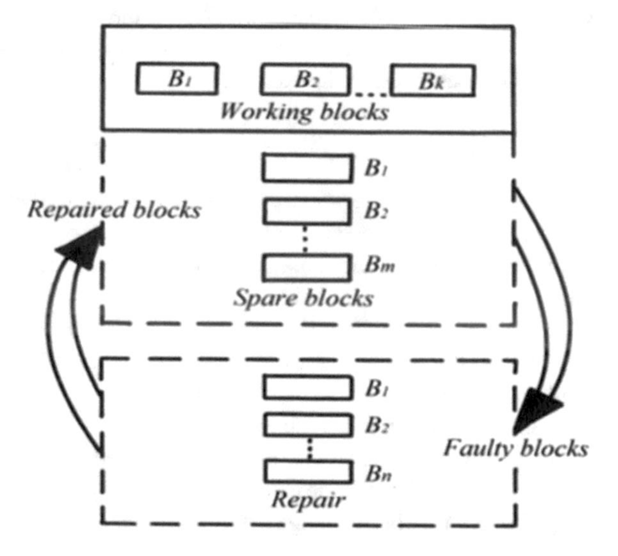

Fig. 3.6 The maintenance and repair system for trains and ships

The average specific cost of outage takes into account the full range of expenses associated with the maintenance and repair of the propulsion system, whose shutdown grounds the vehicle. To simulate this process and to define C_{RAC} for the entire period of operation of the vehicle, we used the method of statistical tests. A more detailed RAC assessment based on the Markov Reward models is presented in Chap. 4.

3.7 Ultimate Results for Icebreaker Ships

To carry out a comparative analysis, hybrid electric ships were selected as the application case. We chose them for investigation for the following reasons:

- The preliminary analysis demonstrated the high competitiveness of the two types of traction motors—induction motor and synchronous motor with permanent magnets—for realizing the ship's electric propulsion systems.
- Restrictions on the weight and dimensions of the traction electric motor of the helicopter allow the use of traction PSM motors only.
- The financial losses due to the wrong choice of traction electric motor for cars and trains are relatively small compared to the losses related to the development and design of the powerful electrical propulsion system installed in ships.
- Due to the high capital cost, high operational cost and heavy expenses incurred in the event of damage, the choice of a traction motor that is not the optimal option for a ship's electric propulsion system can lead to significant financial losses and to a decrease in the ship-owner's profits.
- A large amount of statistical data for the last five years of operation of the traction electric drive used in ships has been collected and systematized.

Thus, two types of ships designated for Arctic navigation—the Amguema-type Arctic ship (Fig. 3.7a) with a diesel-electric propulsion system (Fig. 3.8a) and the diesel-electric icebreaker "Ilya Muromets" (Fig. 3.8a, b)—were considered for evaluation.

We have accepted the sustainable performance of transportation work with minimal cost as a main target function of these two types of ships, as noted in [5]. In this case, for the two types of ships that we compared and which perform the same target function, we must adopt the function "usefulness" with the same value for both compared ship types. In this case, in order to select the best option, we conducted out an analysis of the total cost and total damage over the 25 years of the ship's lifetime [24, 30]. The remaining baseline data is presented in Table 3.8.

(a)

(b)

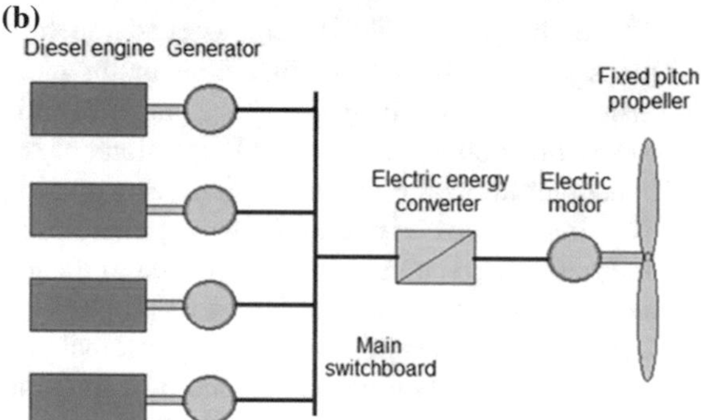

Fig. 3.7 The Amguema-type Arctic ship (**a**) and his propulsion system (**b**)

Given the constantly changing cost of rare-earth metals, different variants of their prices were considered in the analyzing options. The difference in the cost of an IM and a PSM from 10%, up to 50% was taken into consideration as well. The results of the comparative assessment of the full life cycle of an Amguema-type Arctic ship and the icebreaker "Ilya Muromets" are shown in Tables 3.9 and 3.10 respectively. All the parameters in these tables are in millions of euros.

The graphs plotted on the basis of the results of the calculations for the two types of ships are presented, respectively, in Figs. 3.9 and 3.10. The graphs contain the data on an Amguema-type Arctic ship with an IM, which costs 2 million euros (Fig. 3.9), and the data on the icebreaker "Ilya Muromets" with an IM, which costs 3 million euros (Fig. 3.10).

The simulation results show that, for both types of electric propulsion systems, a PSM is the preferred type of traction motor for such ships.

(a)

(b)

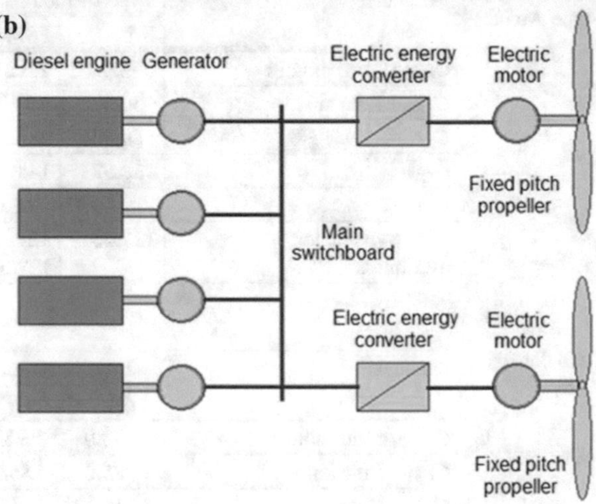

Fig. 3.8 The icebreaker "Ilya Muromets" (**a**) and his propulsion system (**b**)

Table 3.8 Initial Data for Comparison

Indicator	Type of motor			
	Amguema-type Arctic ship		Icebreaker "Ilya Muromets"	
	IM	PSM	IM	PSM
Average efficiency	0.92	0.96	0.92	0.96
Average availability	0.90	0.92	0.90	0.92
Running time rate	0.37	0.37	0.75	0.75
Specific fuel consumption (g/kWh)	260	260	260	260
Full energy generation (GWh)	226.4	217.7	893.4	859.1
Average diesel fuel cost (€/ton)	500	500	500	500
Lifetime fuel consumption (ton)	58,858	56,594	232,254	223,320

Table 3.9 Cost and damage for an Amguema-type Arctic ship

Parameter	Type of motor	
	IM	PSM
C_{CAP} of the motors	2.0	2.2/2.4/3.0
C_{CAP} of the ship	50.0	50.2/50.4/51.0
C_F	29.43	28.23
C_{RAC}	18.57	14.86
C_{EAC}	29.43	28.23
$\sum$ "payment for usefulness"	127.43	121.52/121.72/122.32

Table 3.10 Cost and damage for the icebreaker "Ilya Muromets"

Parameter	Type of motor	
	IM	PSM
C_{CAP} of the motors	3.0	3.3/3.6/4.5
C_{CAP} of the ship	85.0	85.3/85.6/86.5
C_F	116.13	111.66
C_{RAC}	55.71	44.58
C_{EAC}	116.13	111.66
$\sum$ "payment for usefulness"	372.97	353.2/353.5/354.7

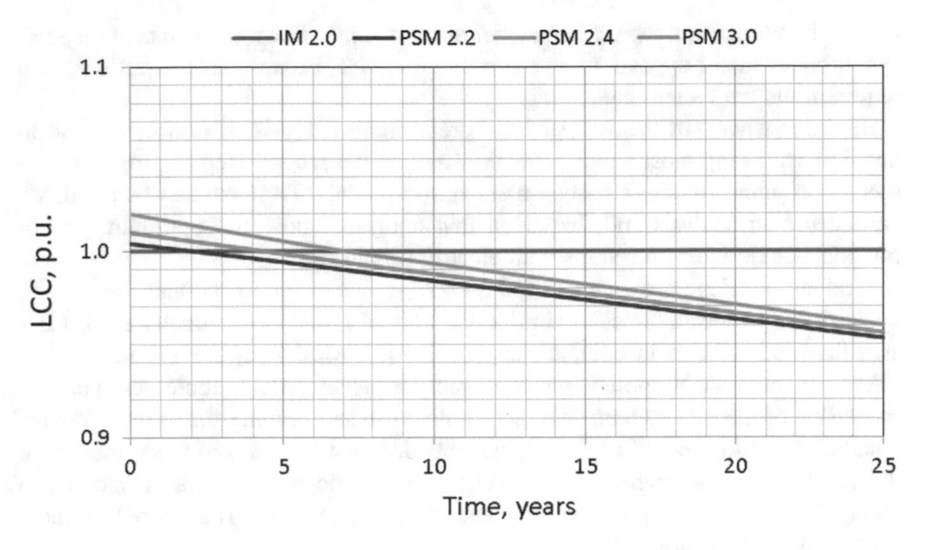

Fig. 3.9 Amguema-type Arctic ship

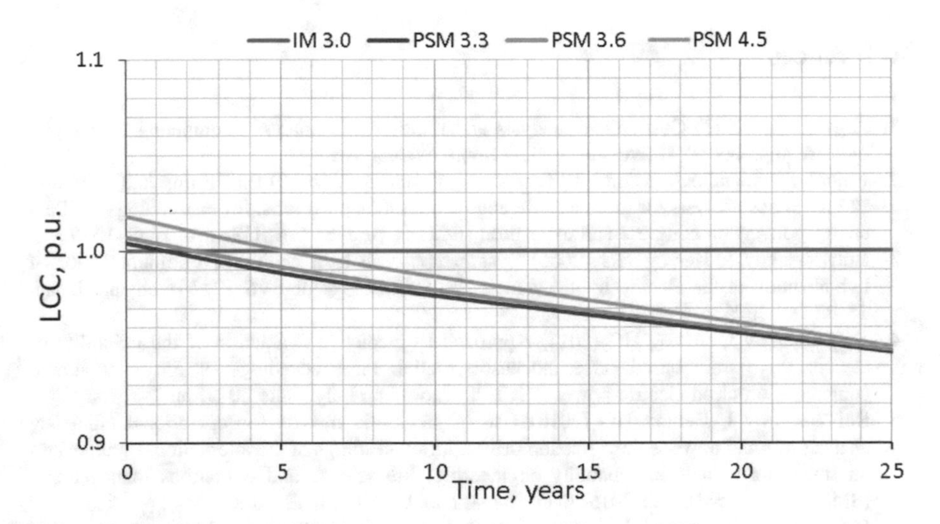

Fig. 3.10 Icebreaker "Ilya Muromets"

3.8 Conclusion

This chapter has presented the methodology of a two-step—preliminary and ultimate—selection of the appropriate electric traction motors for vehicular propulsion systems in different application cases. For some application cases, the results of the preliminary step indicate that there exists more than one adequate electrical machine type. For instance, for cars with specific requirements, it is possible to use any analyzed machine type, whereas, for aircraft applications, a permanent magnets synchronous motor appears to be the optimal choice, because an aircraft's system requirements are rather challenging.

In cases where, in the wake of the preliminary analysis, it is not possible to decide what is the most suitable traction motor, the second step is proposed: the method of refined analysis based on stochastic models. To illustrate this point, we carried out an analysis and, based on that analysis, chose the optimum type of traction electric motor for ships with electric propulsion.

The results of the research showed that, when we take into account the full life cycle of a ship for all considered price variants of rare-earth magnets, a PSM has significant advantages in comparison with an IM. Moreover, the systematic efficiency of the PSM's application as a traction electric motor increases with the increasing of the ship's coefficient of technical use and running time rate. We have concluded that the cost of rare-earth magnets does not have a significant impact on the results of a comprehensive evaluation of electric motor variants and that a change in the cost of magnets by 50% leads to a change in the criterion of "payment for usefulness" by only 0.7%.

References

1. Argyrous G (2011) Cost-benefit analysis and multi-criteria analysis: competing or complementary approaches? University of New South Wales, Sydney
2. Bojoi R, Cavagnino A, Miotto A, Tenconi A, Vaschetto S (2010) Radial flux and axial flux PM machines analysis for more electric engine aircraft applications. In: Proceedings of IEEE energy conversion congress and exposition (ECCE), Atlanta, GA, USA, pp 1672–1679
3. Bolvashenkov I, Herzog HG (2006) System approach to a choice of optimum factor of hybridization of the electric hybrid vehicle. In: Proceedings of EVS 22, Yokohama, Japan, 23–28 Oct 2006, pp 1–12
4. Bolvashenkov I, Herzog HG (2015) Approach to predictive evaluation of the reliability of electric drive train based on a stochastic model. In: Proceedings of 5th international conference on clean electric power (ICCEP), Taormina, Italy, June 2015, pp 1–7
5. Bolvashenkov I, Herzog HG (2016) Use of stochastic models for operational efficiency analysis of multi power source traction drives. In: Proceedings of the international symposium on stochastic models in reliability engineering, life science and operations management, (SMRLO'16), 15–18 Feb 2016, Beer Sheva, Israel, 2016, pp 124–130
6. Bolvashenkov I, Herzog HG, Rubinraut A, Romanovskiy V (2014) Possible ways to improve the efficiency and competitiveness of modern ships with electric propulsion systems. In Proceedings of 10th IEEE vehicle power and propulsion conference (VPPC), Coimbra, Portugal, November 2014, pp 1–9
7. Bolvashenkov I, Kammermann HJ, Herzog HG (2016) Methodology for selecting electric traction motors and its application to vehicle propulsion systems. In: Proceedings of international symposium on power electronics, electrical drives, automation and motion (SPEEDAM), 22–24 June 2016, Anacapri, Italy, pp 1214–1219
8. Bolvashenkov I, Kammermann HJ, Herzog HG (2016) Research on reliability and fault tolerance of traction multi-phase permanent magnet synchronous motors based on Markov-models for multi-state systems. In: Proceedings of international symposium on

power electronics, electrical drives, automation and motion, (SPEEDAM), 22–24 June 2016, Anacapri, Italy, pp 1166–1171

9. Bolvashenkov I, Kammermann HJ, Willerich S, Herzog HG (2015) Comparative study for the optimal choice of electric traction motors for a helicopter drive train. In: Proceedings of the 10th conference on sustainable development of energy, water and environment systems (SDEWES), 27 Sept–3 Oct 2015, Dubrovnik, Croatia, 2015, pp 1–9
10. Brahman T (1984) Multicriteriality and choice of the alternative in technique. Radio and Communications, Moscow (in Russian)
11. Buecherl D, Bolvashenkov I, Herzog HG (2009) Verification of the optimum hybridization factor as design parameter of hybrid electric vehicles. In: Proceedings of 5th IEEE vehicle power and propulsion conference (VPPC'09), 7–11 Sept 2009, Dearborn, Michigan, USA, 2009, pp 847–851
12. Dajaku G, Gerling D (2012) Efficiency improvements of electric machines for automotive application. In: Proceedings of 26th congress EVS, (EVS-26), Los Angeles, California, May 2012, pp 1–7
13. Dhillon BS (2010) Life cycle costing for engineers. Taylor and Francis Group, New York
14. Duffy MC (1992) Three-phase motor in railway traction. IEEE Proc Sci Meas Technol 139 (6):329–337
15. Ermolin NP, Zerichin IP (1981) Zuverlässigkeit elektrischer maschinen. Verl. Technik, Berlin (in German)
16. Frenkel I, Bolvashenkov I, Herzog HG, Khvatskin L (2016) Performance availability assessment of combined multi power source traction drive considering real operational conditions. Transp Telecommun 17(3):179–191
17. Frenkel I, Bolvashenkov I, Herzog HG, Khvatskin L (2017) Operational sustainability assessment of multipower source traction drive. In: Mathematics applied to engineering. Elsevier, pp 191–203
18. Jack AG, Mecrow BC, Haylock JA (1996) A comparative study of permanent magnet and switched reluctance motors for high-performance fault-tolerant applications. IEEE Trans Ind Appl 32(4):889–895
19. Kammermann J, Bolvashenkov I, Herzog HG (2015) Approach for comparative analysis of electric traction machines. In: Proceedings of 3rd international conference on electrical systems for aircraft, railway, ship propulsion, and road vehicles (ESARS) Aachen, Germany, Mar 2015, pp 1–5
20. Kammermann J, Bolvashenkov I, Herzog HG (2017) Reliability of induction machines: statistics, tendencies, and perspectives. In: Proceedings of 26th IEEE international symposium on industrial electronics (ISIE), 19th–21th June 2017, Edinburgh, UK, 2017, pp 1843–1847
21. Lauger E (1982) Reliability in electrical and electronic components and systems. North-Holland, Amsterdam
22. Lebsir A, Bentounsi A, Rebbah R, Belakehal S, Benbouzid MEH (2013) Comparative study of PMSM and SRM capabilities. In: Proceedings of 4th international conference on power engineering, energy and electrical drives (IEEE POWERENG13), Istanbul, Turkey, 2013, pp 760–763
23. Mahdavi S, Herold T, Hameyer K (2013) Thermal modelling as a tool to determine the overload capability of electrical machines. In: Proceedings of international conference on electrical machines and systems (ICEMS), Busan, Korea, Oct 2013, pp 454–458
24. Müller C, Vogt K, Ponick B (2008) Berechnung elektrischer maschinen. Wiley-VCH Verlag GmbH & Co. KGaA, Weinheim (in German)
25. Pearce D, Atkinson G, Mourato S (2006) Cost-benefit analysis and the environment. OECD Publishing, Paris
26. Schramm A, Gerling D (2006) Researches on the suitability of switched reluctance machines and permanent magnet machines for specific aerospace applications demanding fault tolerance. In: Proceedings of SPEEDAM, May 2006, pp 56–60
27. Spyropoulos DV, Mitronikas ED (2013) A review on the faults of electric machines used in electric ships. In: Advances in power electronics. Hindawi Publishing Corporation, pp 1–8

28. Zeraoulia M, Benbouzid MEH, Diallo D (2006) Electric motor drive selection issues for HEV propulsion systems: a comparative study. IEEE Trans Veh Technol 55(6):1756–1764
29. Zopounidis C, Pardalos PM (2010) Handbook of multicriteria analysis, applied optimization. Springer, Berlin
30. Zvonov VA, Kozlov AV, Kutenev VF (2001) Environmental safety of the car in a full life cycle. NAMI, Moscow (in Russian)

Chapter 4
The Markov Reward Approach for Selecting a Traction Electric Motor Based on Reliability Features

Abstract This chapter presents the Markov reward approach to the comparative analysis of different types of traction electric motors for hybrid-electric propulsion systems, that is, multi-state systems (MSSs), for icebreaker ships operating in Arctic waters. The preliminary results show that there are several appropriate machine types and that it is therefore necessary to define in advance what kind of equipment would be the most appropriate in order to arrive at a proper decision on this matter. In this chapter, we discuss how we developed the Markov reward approach for computing the MSS's average availability, average converted power and reliability-associated cost, all the while taking into consideration the vehicular operational conditions of electric motors. We also propose what, in our opinion, is the optimal type of traction electric motor.

Keywords Electrical propulsion system · Traction motor · Markov reward model
Multi-state system (MSS) · Average availability · Average converted power
Reliability-associated cost

4.1 Introduction

The design and construction of many types of hybrid-electric propulsion systems for ships constitute a widely discussed subject in the professional literature [1–3, 10]. Since the usage and requirements differ from one type of ship to another, the present chapter offers a comparison of different traction electric motors (TEM), the selection of which will depend on the specific type of ship for which we are choosing the TEM and on the operating conditions, requirements and limitations of these various motors.

For the purpose of our investigation, we have decided that the ship in which we want to install a TEM is an Amguema-type ship with a diesel-electric propulsion system that is used for Arctic navigation (Fig. 4.1) [4, 6]. A detailed description of the Amguema-type ship was presented in Chap. 1.

I. Bolvashenkov et al., *Safety-Critical Electrical Drives*, SpringerBriefs in Electrical and Computer Engineering, https://doi.org/10.1007/978-3-319-89969-5_4

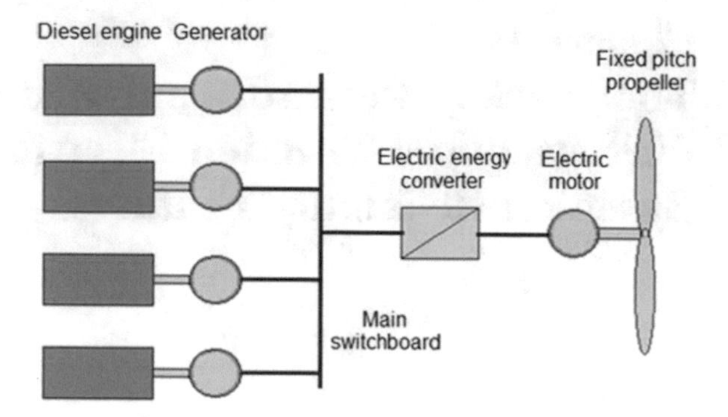

Fig. 4.1 An Amguema-type ship's diesel-electric propulsion system

As noted in [5, 8, 15–19], fuel costs make up 70% of a ship's operating costs; in the selection of a traction electric motor, efficiency is a highly important consideration. In terms of efficiency, the most promising machine types for marine electric propulsion systems are the Induction Machine (IM) and the Permanent Magnet Synchronous Machine (PSM).

The following factors must be weighed in a comparative analysis: the multi-state system's average availability, converted power and reliability-associated cost. In order to conduct the comparative analysis, we adopted the Markov reward approach.

4.2 The Markov Reward Approach for Calculating the Reliability Measures of a Multi-state Systems

The Markov reward approach was first introduced by Howard [9]. In [7, 11–14], one can find a detailed depiction of Markov reward models and descriptions of their employment in the reliability analysis of multi-state systems (MSSs).

We will assume that the Markov model for the multi-state system has K states, which can be represented by a state-space diagram, as well as transitions between different states. Intensities $a_{ij}, i,j = 1, \ldots, K$ of transitions from state i to state j are defined in terms of failure and repair rates. For an aging system, the failure rate $\lambda(t)$ increases with age. In the case of minimal repairs, the intensities of transitions from state i to state j are dependent on time.

We can further assume that, while the system is in any state i during any time unit, some payment r_{ii} will be made. We can also assume that, if there is a transition from state i to state j, the amount r_{ij} will be paid for each transition. The amounts r_{ii} and r_{ij} are rewards and can be presented as a matrix of rewards $\boldsymbol{r}$.

The objective here is to compute the total expected reward accumulated from $t = 0$, when the system begins its evolution in the state space, up to the time $t = T$ under specified initial conditions.

Let us define the total expected reward accumulated up to time t as $V_j(t)$, if the system begins its evolution at time $t = 0$ from state j. According to Howard [9], the following system of differential equations must be solved in order to find this total expected reward:

$$\frac{dV_j(t)}{dt} = r_{jj} + \sum_{\substack{i=1 \\ i \neq j}}^{K} a_{ij} r_{ij} + \sum_{i=1}^{K} a_{ij} V_i(t), \quad j = 1, 2, \ldots, K \tag{4.1}$$

The system (4.1) should be solved under initial conditions: $V_j(0) = 0, j = 1, 2, \ldots, K$.

The total expected reward can be found from differential equations (4.1) through the substitution of formulae for the rate of failures $\lambda(t)$ and the rate of repairs μ instead of the corresponding a_{ij} values.

4.2.1 The Computation of Rewards for the MSS's Average Availability

In order to compute availability, we partitioned the set of states $\mathbf{g}$ into $\mathbf{g}_0$, the set of operational or acceptable system states, and $\mathbf{g}_f$, the set of failed or unacceptable states. The acceptability of system states depends on the relation between the MSS's output performance and the desired level of this performance—a demand that is determined outside the system. In general, demand $W(t)$ is also a random process that can take discrete values from the set $\mathbf{w} = \{w_1, \ldots, w_M\}$. The desired relation between a system's performance and the demand at any time t can be expressed by the acceptability function $\Phi(G(t), W(t))$. The acceptable system states can be expressed as $\Phi(G(t), W(t)) \geq 0$, while the unacceptable states can be expressed as $\Phi(G(t), W(t)) < 0$. Often the MSS's performance will be equal to, or will exceed, the demand. In such cases, the acceptability function can be expressed as $\Phi(G(t), W(t)) = G(t) - W(t)$, while the criterion of state acceptability can be expressed as $\Phi(G(t), W(t)) = G(t) - W(t) \geq 0$.

If we assume that the required demand level is constant $W(t) \equiv w$, the criterion of state acceptability can be expressed as $\Phi(G(t), w) = G(t) - w \geq 0$. Thus, all system states with a level of performance that is greater than, or equal to, w correspond to the set of acceptable states, while all system states with a level of performance that is lower than w correspond to the set of unacceptable states.

The MSS's instantaneous (point) availability $A(t)$ is the probability that the MSS at instant $t > 0$ is in one of the acceptable states:

$$A(t) = \sum_{G(t) \geq W(t)} P_i(t) \tag{4.2}$$

where $P_i(t)$ is the probability that at instant t the system is in state i.

The MSS's average availability $\overline{A}(T)$ is defined as a mean fraction of time when the system resides in the set of acceptable states during the time interval [0, T],

$$\bar{A}(T) = \frac{1}{T}\int_0^T A(t)dt \tag{4.3}$$

To assess $\overline{A}(T)$ for an MSS, the rewards in matrix $\boldsymbol{r}$ should be determined in the following manner:

- The rewards associated with all acceptable states should be defined as 1.
- The rewards associated with all unacceptable states as well as all the rewards associated with all transitions should be defined as zero.

The total expected reward $V_j(T)$ accumulated during interval [0, T] defines a time when the MSS will be in the set of acceptable states when state j is the initial state. This reward should be regarded as the solution of the system (4.1). After we have solved the system (4.1) and found $V_j(t)$, the MSS's average availability can be obtained for every $j = 1, \ldots, K$:

$$\overline{A}_j(T) = V_j(T)/T \tag{4.4}$$

Usually the first state is determined as an initial state and

$$\overline{A}(T) = V_1(T)/T \tag{4.5}$$

4.2.2 The Computation of Rewards for the MSS's Average Converted Power

The MSS's instantaneous (point) converted power $E(t)$ is calculated as a multiplication of performance given the probability that the MSS at instant $t > 0$ is in one of the acceptable states:

$$E(t) = \sum_{G(t) > 0} g_i P_i(t) \tag{4.6}$$

where $P_i(t)$ is the probability that at instant t the system is in state i and g_i is the level of performance in state i.

The MSS's average converted power $\overline{E}(T)$ is defined as follows:

$$\overline{E}(t) = \int_0^T E(t)dt \tag{4.7}$$

To assess $\overline{E}(t)$ for an MSS, the rewards in matrix $\boldsymbol{r}$ can be determined in the following manner:

- The rewards associated with all acceptable states should be defined as g_i.
- The rewards associated with all unacceptable states as well as all the rewards associated with all transitions should be defined as zero.

The total expected reward $V_j(T)$ accumulated during interval [0, *T*] defines a total converted power that the MSS will convert during being in the set of acceptable states where state *j* is the initial state. This reward should be regarded as the solution of the system (4.1). After we have solved the system (4.1) and found $V_j(t)$, the MSS's average converted power can be obtained for every $j = 1, \ldots, K$:

$$\overline{E}_i(T) = V_i(T)/T \tag{4.8}$$

Usually the first state is determined as an initial state and

$$\overline{E}(T) = V_1(T)/T \tag{4.9}$$

4.2.3 The Computation of Rewards for the MSS's Reliability-Associated Cost

We define the MSS's reliability-associated cost (*RAC*) as the total cost incurred by the user in operating and maintenance the system during its lifetime. Therefore,

$$RAC = C_{OC} + C_r + C_p \tag{4.10}$$

where

- C_{op} is the system's operating cost accumulated during the system's lifetime;
- C_r is the repair cost incurred by the user in repairing and maintaining the system during its lifetime;
- C_p is the penalty cost, which is accumulated during the system's lifetime and which is paid when the system fails.

Let *T* be the system's lifetime. During this time, the system may be in an acceptable state (system functioning) or in an unacceptable one (system failure). After any failure, a corresponding repair action is performed and the system returns to one of the previously acceptable states.

To assess the *RAC* for an MSS, the rewards in matrix ***r*** can be determined in the following manner:

- The rewards associated with all acceptable states should be defined as C_{op}.
- The rewards associated with all unacceptable states should be defined as C_p.
- The rewards associated with transitions, defined repairs and maintenance should be defined as C_r.

The total expected reward $V_j(T)$ accumulated during interval [0, *T*] defines the *RAC* as the total cost incurred by the user in operating and maintenance the system during its lifetime. This reward should be regarded as a solution of the system (4.1). After we have solved the system (4.1) and found $V_j(t)$, the MSS's average availability can be obtained for every $j = 1, \ldots, K$:

$$RAC = V_1(T) \tag{4.11}$$

4.3 The Calculation of Reliability and the Comparison of Different Types of Traction Electric Motors

4.3.1 The Calculation of the MSS's Average Availability for Constant and Seasonal Stochastic Demand

Let us consider a system consisting of two traction electric motors (Fig. 4.2) with a nominal converted power to the propeller of 2 × 2,750 = 5,500 kW as a multi-state system. This system's state-space diagrams (Fig. 4.3a, b) have three different performance levels: perfect functioning ($g_1 = 5{,}500$ kW), reduced power ($g_{2,3} = 2{,}750$ kW), and total failure ($g_4 = 0$). The failure rates of each motor ($\lambda_1(t)$ and $\lambda_2(t)$) are increasing functions and indicate the aging of the motor. The repair rates for each motor are the same and are equal to μ. Only one repair equipment is available; thus, the repair rate when both motors are failed is $1/2\mu$.

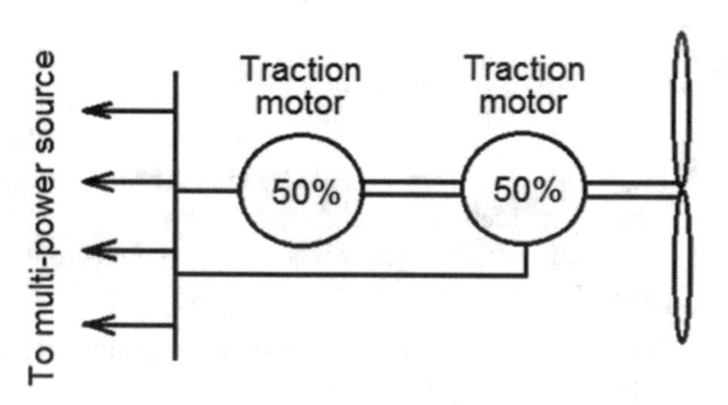

Fig. 4.2 Traction electric motor

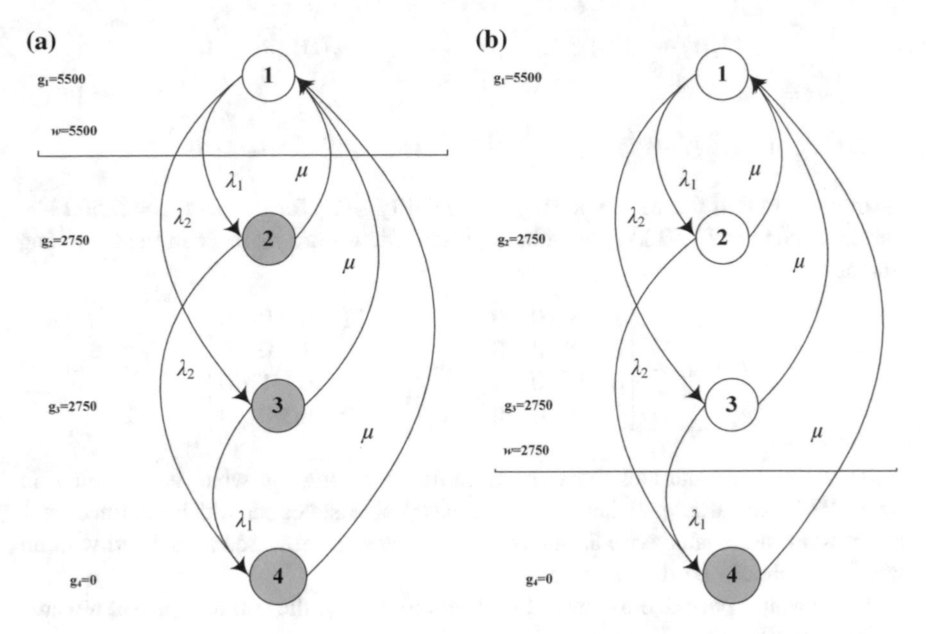

Fig. 4.3 State-space diagram of the system for demand **a** w = 5,500 kW, **b** w = 2,750 kW

The required converted power to the propeller (the demand) is defined by the weather conditions. The high-level demand in winter period is w = 5,500 kW. The low-level (summer period) demand is w = 2,750 kW. Motor system failure is defined as the system's converted power falling below demand level w.

The state-space diagrams for a system consisting of two traction electric motors designed to meet winter period demand w = 5,500 kW and summer period demand w = 2,750 kW are presented in Fig. 4.3a, b.

As noted above, in the winter period, the demand is w = 5,500 kW. Only the first state, where the system's converted power is g_1 = 5,500 kW, is the acceptable state ($g_1 - w = 5500 - 5500 = 0$), while, in rest states, the criterion of the state's acceptability is negative ($g_{2,3} - w = 2750 - 5500 \leq 0$, $g_4 - w = 0 - 5500 \leq 0$). Thus, states 2, 3 and 4 are unacceptable states.

In the summer period, demand is w = 2,750 kW. The system's converted power is represented as g_1 = 5,500 kW, $g_{2,3}$ = 2,750 kW. According to the criterion of state acceptability ($g_1 - w = 5500 - 2750 \geq 0$, $g_{2,3} - w = 2750 - 2750 = 0$), only state 4 is an unacceptable state.

The unacceptable states are shaded in grey.

As indicated in the state-space diagrams in Fig. 4.3a, b, the transition intensity matrix $\boldsymbol{a}$ can be presented as follows:

$$a = \begin{vmatrix} -(\lambda_1(t) + \lambda_2(t)) & \lambda_1(t) & \lambda_2(t) & 0 \\ \mu & -(\lambda_2(t) + \mu) & 0 & \lambda_2(t) \\ \mu & 0 & -(\lambda_1(t) + \mu) & \lambda_1(t) \\ 1/2\mu & 0 & 0 & -\mu \end{vmatrix} \quad (4.12)$$

In order to find the MSS's average availability $\bar{A}(T)$ for demand w = 5,500 kW and demand w = 2,750 kW, we should present the reward matrix $\boldsymbol{r}$ in the following manner:

$$\boldsymbol{r}_{5500} = \begin{vmatrix} 1 & 0 & 0 & 0 \\ 0 & 0 & 0 & 0 \\ 0 & 0 & 0 & 0 \\ 0 & 0 & 0 & 0 \end{vmatrix} \quad \boldsymbol{r}_{2750} = \begin{vmatrix} 1 & 0 & 0 & 0 \\ 0 & 1 & 0 & 0 \\ 0 & 0 & 1 & 0 \\ 0 & 0 & 0 & 0 \end{vmatrix} \quad (4.13)$$

These reward matrixes have been built according to what is specified in Sect. 4.2.1. Rewards associated with all acceptable states should be defined as 1 and rewards associated with all unacceptable states and rewards associated with all transitions should be defined as zero.

In the winter period, only reward r_{11} is equal to 1. In the summer period rewards $r_{11} = r_{22} = r_{33} = 1$.

The systems of differential equations must be solved for the transition intensity matrix (4.12) and the reward matrix (4.13) under initial conditions $V_j(0) = 0,\ j = 1,2,3,4$.

The system of differential equations for demand w = 5,500 kW is as follows:

$$\begin{aligned} \frac{dV_1(t)}{dt} &= 1 - (\lambda_1(t) + \lambda_2(t)) \cdot V_1(t) + \lambda_1(t) \cdot V_2(t) + \lambda_2(t) \cdot V_3(t) \\ \frac{dV_2(t)}{dt} &= \mu \cdot V_1(t) - (\lambda_2(t) + \mu) \cdot V_2(t) + \lambda_2(t) \cdot V_4(t) \\ \frac{dV_3(t)}{dt} &= \mu \cdot V_1(t) - (\lambda_1(t) + \mu) \cdot V_3(t) + \lambda_1(t) \cdot V_4(t) \\ \frac{dV_4(t)}{dt} &= 1/2\mu \cdot V_1(t) - 1/2\mu \cdot V_4(t) \end{aligned} \quad (4.14)$$

The system of differential equations for demand w = 2,750 kW is as follows:

$$\begin{aligned} \frac{dV_1(t)}{dt} &= 1 - (\lambda_1(t) + \lambda_2(t)) \cdot V_1(t) + \lambda_1(t) \cdot V_2(t) + \lambda_2(t) \cdot V_3(t) \\ \frac{dV_2(t)}{dt} &= 1 + \mu \cdot V_1(t) - (\lambda_2(t) + \mu) \cdot V_2(t) + \lambda_2(t) \cdot V_4(t) \\ \frac{dV_3(t)}{dt} &= 1 + \mu \cdot V_1(t) - (\lambda_1(t) + \mu) \cdot V_3(t) + \lambda_1(t) \cdot V_4(t) \\ \frac{dV_4(t)}{dt} &= 1/2\mu \cdot V_1(t) - 1/2\mu \cdot V_4(t) \end{aligned} \quad (4.15)$$

After we have solved the systems (4.14)–(4.15) and found $V_j(t)$, the MSS's average availability can be expressed as $\overline{A}(T) = V_1(T)/T$.

In order to take into consideration the impact of seasonal stochastic demand on our multi-state model, we incorporated the two-state Markov process $W(t)$ with minimum converted power level w = 2,750 kW and maximum converted power level w = 5,500 kW in the state-space diagrams in Fig. 4.3a, b. The new state-space diagram for the Markov process is presented in Fig. 4.4.

States 1–4 represent the winter period, states 5–8 the summer period. In the winter period, only state 1 is acceptable, while, in the summer period, states 1–3 are acceptable. Unacceptable states are shaded in grey.

According to the state-space diagram in Fig. 4.4, the following transition intensity matrix $\boldsymbol{a}$ can be represented as follows:

$$\boldsymbol{a} = \begin{pmatrix} a_{11} & \lambda_1(t) & \lambda_2(t) & 0 & \lambda_w & 0 & 0 & 0 \\ \mu & a_{22} & 0 & \lambda_2(t) & 0 & \lambda_w & 0 & 0 \\ \mu & 0 & a_{33} & \lambda_1(t) & 0 & 0 & \lambda_w & 0 \\ 1/2\mu & 0 & 0 & a_{44} & 0 & 0 & 0 & \lambda_w \\ \lambda_s & 0 & 0 & 0 & a_{55} & \lambda_1(t) & \lambda_2(t) & 0 \\ 0 & \lambda_s & 0 & 0 & \mu & a_{66} & 0 & \lambda_2(t) \\ 0 & 0 & \lambda_s\cdot & 0 & \mu & 0 & a_{77} & \lambda_1(t) \\ 0 & 0 & 0 & \lambda_s & 1/2\mu & 0 & 0 & a_{88} \end{pmatrix} \quad (4.16)$$

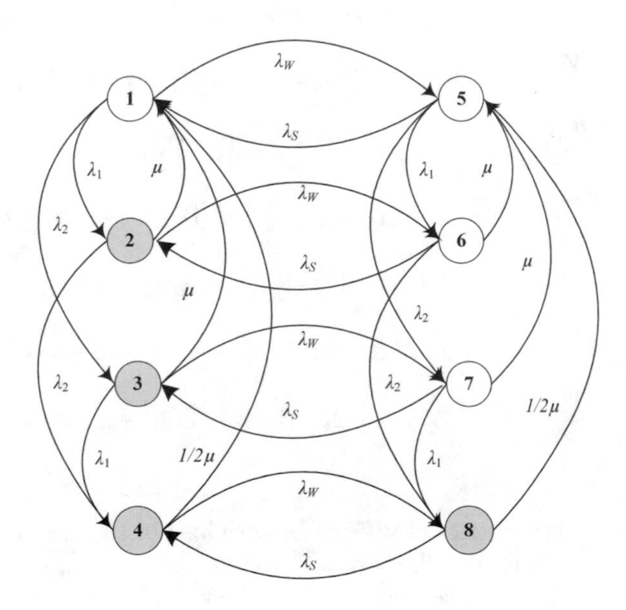

Fig. 4.4 State-space diagram of a system with seasonal stochastic demand

where

$$\begin{aligned} a_{11} &= -(\lambda_1(t) + \lambda_2(t) + \lambda_w) & a_{55} &= -(\lambda_1(t) + \lambda_2(t) + \lambda_s) \\ a_{22} &= -(\lambda_2(t) + \mu + \lambda_w) & a_{66} &= -(\lambda_2(t) + \mu + \lambda_s) \\ a_{33} &= -(\lambda_1(t) + \mu + \lambda_w) & a_{77} &= -(\lambda_1(t) + \mu + \lambda_s) \\ a_{44} &= -(1/2\mu + \lambda_w) & a_{88} &= -(1/2\mu + \lambda_s) \end{aligned}$$

The reward matrix $\boldsymbol{r}$ must be represented in the following manner:

$$\boldsymbol{r} = \begin{vmatrix} 1 & 0 & 0 & 0 & 0 & 0 & 0 & 0 \\ 0 & 0 & 0 & 0 & 0 & 0 & 0 & 0 \\ 0 & 0 & 0 & 0 & 0 & 0 & 0 & 0 \\ 0 & 0 & 0 & 0 & 0 & 0 & 0 & 0 \\ 0 & 0 & 0 & 0 & 1 & 0 & 0 & 0 \\ 0 & 0 & 0 & 0 & 0 & 1 & 0 & 0 \\ 0 & 0 & 0 & 0 & 0 & 0 & 1 & 0 \\ 0 & 0 & 0 & 0 & 0 & 0 & 0 & 0 \end{vmatrix} \tag{4.17}$$

The system of differential equations with seasonal stochastic demand can be expressed as follows:

$$\begin{aligned} \frac{dV_1(t)}{dt} &= 1 - (\lambda_1(t) + \lambda_2(t) + \lambda_w) \cdot V_1(t) + \lambda_1(t) \cdot V_2(t) + \lambda_2(t) \cdot V_3(t) + \lambda_w \cdot V_5(t) \\ \frac{dV_2(t)}{dt} &= \mu \cdot V_1(t) - (\lambda_2(t) + \mu + \lambda_w) \cdot V_2(t) + \lambda_2(t) \cdot V_4(t) + \lambda_w \cdot V_6(t) \\ \frac{dV_3(t)}{dt} &= \mu \cdot V_1(t) - (\lambda_1(t) + \mu + \lambda_w) \cdot V_3(t) + \lambda_1(t) \cdot V_4(t) + \lambda_w \cdot V_7(t) \\ \frac{dV_4(t)}{dt} &= 1/2\mu \cdot V_1(t) - (1/2\mu + \lambda_w) \cdot V_4(t) + \lambda_w \cdot V_8(t) \\ \frac{dV_5(t)}{dt} &= 1 + \lambda_s \cdot V_1(t) - (\lambda_1(t) + \lambda_2(t) + \lambda_s) \cdot V_5(t) + \lambda_1(t) \cdot V_6(t) + \lambda_2(t) \cdot V_7(t) \\ \frac{dV_6(t)}{dt} &= 1 + \lambda_s \cdot V_2(t) + \mu \cdot V_5(t) - (\lambda_2(t) + \mu + \lambda_s) \cdot V_6(t) + \lambda_2(t) \cdot V_8(t) \\ \frac{dV_7(t)}{dt} &= 1 + \lambda_s \cdot V_3(t) + \mu \cdot V_5(t) - (\lambda_1(t) + \mu + \lambda_s) \cdot V_7(t) + \lambda_1(t) \cdot V_8(t) \\ \frac{dV_8(t)}{dt} &= \lambda_s \cdot V_4(t) + 1/2\mu \cdot V_5(t) - (1/2\mu + \lambda_s) \cdot V_8(t) \end{aligned} \tag{4.18}$$

The systems of differential equations (4.18) must be solved under initial conditions $V_j(0) = 0,\ j = 1, 2, \ldots, 8$.

After we have solved the system (4.18) and found $V_j(t)$, the MSS's average availability can be expressed as $\overline{A}(T) = V_1(T)/T$.

We performed calculations using the failure and repair rates indicated below.

The failure rate of each IM motor is $\lambda^{\mathrm{IM}}(t) = 0.92 + 0.05t$ year^{-1}. The failure rate of each PSM motor is $\lambda^{\mathrm{PSM}}(t) = 0.39 + 0.05t$ year^{-1}. The repair rate for both kinds of motors is $\mu = 60.8$ year^{-1}.

The results of the calculation for one year and for a 25-year period are presented in Figs. 4.5 and 4.6.

The calculation of availability indicates that a PSM motor is preferable; its availability is much greater than that of an IM motor. As one can see, the average availability curve is a decreasing function for an aging system.

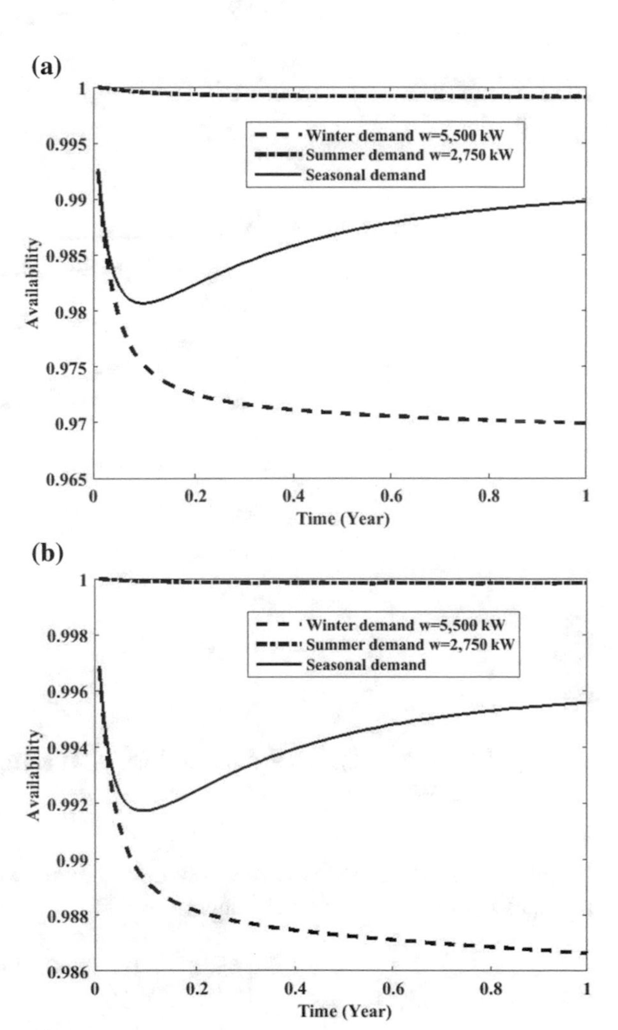

Fig. 4.5 Comparison of the MSS's average availability with an IM motor (**a**) and a PSM motor (**b**) for one-year period

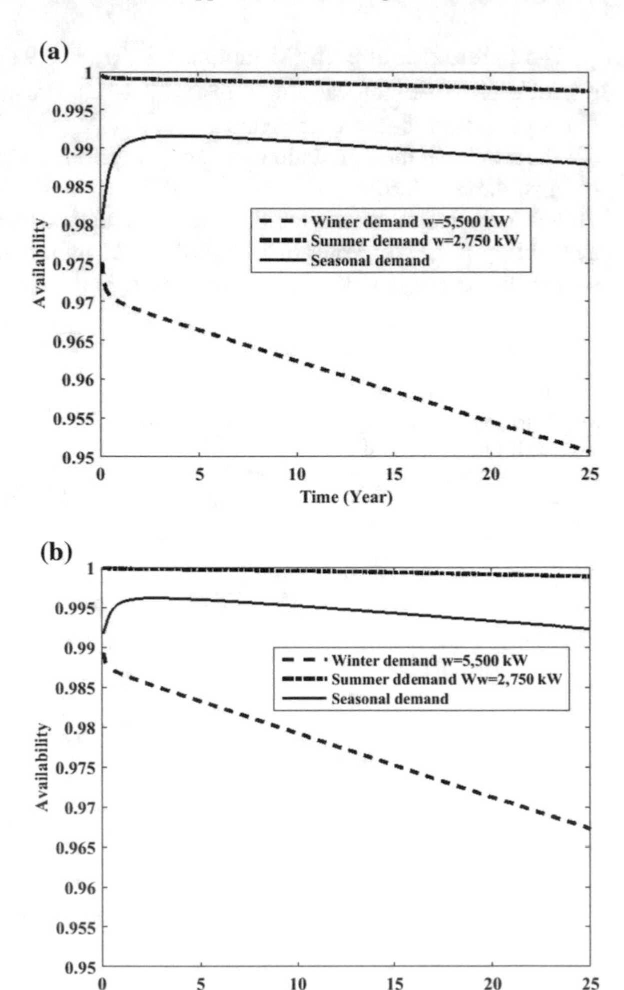

Fig. 4.6 Comparison of the MSS's average availability with an IM motor (**a**) and a PSM motor (**b**) for a 25-year period

4.3.2 *The Calculation of the MSS's Average Converted Power*

In order to find the MSS's average converted power, we should present the reward matrixes $\boldsymbol{r_{Perf}}$ in the following manner:

$$\boldsymbol{r_{Perf}} = \begin{vmatrix} 5500 & 0 & 0 & 0 \\ 0 & 2750 & 0 & 0 \\ 0 & 0 & 2750 & 0 \\ 0 & 0 & 0 & 0 \end{vmatrix} \tag{4.19}$$

This reward matrix has been built according to what is specified in Sect. 4.2.2. Rewards associated with all acceptable states should be defined as being associated with three levels of converted power: $r_{11} = 5500$, $r_{22} = 2750$ and $r_{33} = 2750$; rewards associated with all unacceptable states and rewards associated with all transitions should be defined as zero.

The system of differential equations must be solved for the transition intensity matrix (4.12) and the reward matrix (4.19) under initial conditions $V_j(0) = 0, j = 1, 2, \ldots, 4$.

The system of differential equations is as follows:

$$\begin{aligned}
\frac{dV_1(t)}{dt} &= 5500 - (\lambda_1(t) + \lambda_2(t)) \cdot V_1(t) + \lambda_1(t) \cdot V_2(t) + \lambda_2(t) \cdot V_3(t) \\
\frac{dV_2(t)}{dt} &= 2750 + \mu \cdot V_1(t) - (\lambda_2(t) + \mu) \cdot V_2(t) + \lambda_2(t) \cdot V_4(t) \\
\frac{dV_3(t)}{dt} &= 2750 + \mu \cdot V_1(t) - (\lambda_1(t) + \mu) \cdot V_3(t) + \lambda_1(t) \cdot V_4(t) \\
\frac{dV_4(t)}{dt} &= 1/2\mu \cdot V_1(t) - 1/2\mu \cdot V_4(t)
\end{aligned} \tag{4.20}$$

After we have solved the system and found $V_j(t)$, the MSS's average converted power can be expressed as $\overline{E}(T) = V_1(T)/T$. The results of the calculation are presented in Fig. 4.7. Here one can see the advantages of a PSM motor in comparison with an IM motor and one can also see the expected reduction of converted power that will depend on the aging of both motors.

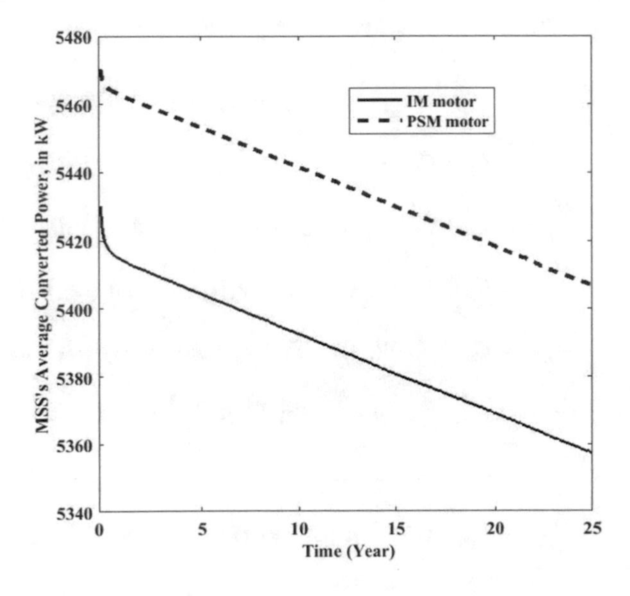

Fig. 4.7 The MSS's average converted power

4.3.3 The Calculation of the MSS's Reliability-Associated Cost

In order to find the MSS's reliability-associated cost, we need to take into consideration the impact of seasonal stochastic demand on our multi-state model and we should present the reward matrixes $\boldsymbol{r_{RAC}}$ in the following manner:

$$r_{RAC} = \begin{pmatrix} C_{op} & 0 & 0 & 0 & 0 & 0 & 0 & 0 \\ C_r & C_p & 0 & 0 & 0 & 0 & 0 & 0 \\ C_r & 0 & C_p & 0 & 0 & 0 & 0 & 0 \\ C_r & 0 & 0 & C_p & 0 & 0 & 0 & 0 \\ 0 & 0 & 0 & 0 & C_{op} & 0 & 0 & 0 \\ 0 & 0 & 0 & 0 & C_r & C_{op} & 0 & 0 \\ 0 & 0 & 0 & 0 & C_r & 0 & C_{op} & 0 \\ 0 & 0 & 0 & 0 & C_r & 0 & 0 & C_p \end{pmatrix} \quad (4.21)$$

This reward matrix has been built according to what is specified in Sect. 4.2.3.

- Rewards associated with all acceptable states should be defined as the operating cost ($r_{11} = r_{55} = r_{66} = r_{77} = C_{op}$).
- Rewards associated with all unacceptable states should be defined as the penalty cost ($r_{22} = r_{33} = r_{44} = r_{88} = C_p$).
- Rewards associated with all transitions, repairs and maintenance should be defined as the repair cost ($r_{21} = r_{31} = r_{41} = r_{65} == r_{75} = r_{85} = C_p$).
- All other rewards should be defined as zero.

The system of differential equations (4.1) must be solved for the reward matrix (4.21) under initial conditions $V_j(0) = 0, j = 1, 2, \ldots, 8$.

The system of differential equations is as follows:

$$\begin{aligned}
\frac{dV_1(t)}{dt} &= C_{op} - (\lambda_1 + \lambda_2 + \lambda_w) \cdot V_1(t) + \lambda_1 \cdot V_2(t) + \lambda_2 \cdot V_3(t) + \lambda_w \cdot V_5(t) \\
\frac{dV_2(t)}{dt} &= C_p + C_r \cdot \mu + \mu \cdot V_1(t) - (\lambda_2 + \mu + \lambda_w) \cdot V_2(t) + \lambda_2 \cdot V_4(t) + \lambda_w \cdot V_6(t) \\
\frac{dV_3(t)}{dt} &= C_p + C_r \cdot \mu + \mu \cdot V_1(t) - (\lambda_1 + \mu + \lambda_w) \cdot V_3(t) + \lambda_1 \cdot V_4(t) + \lambda_w \cdot V_7(t) \\
\frac{dV_4(t)}{dt} &= C_p + 1/2C_r \cdot \mu + 1/2\mu \cdot V_1(t) - (1/2\mu + \lambda_w) \cdot V_4(t) + \lambda_w \cdot V_8(t) \\
\frac{dV_5(t)}{dt} &= C_{op} + \lambda_s \cdot V_1(t) - (\lambda_1 + \lambda_2 + \lambda_s) \cdot V_5(t) + \lambda_1 \cdot V_6(t) + \lambda_2 \cdot V_7(t) \\
\frac{dV_6(t)}{dt} &= C_{op} + C_r \cdot \mu + \lambda_s \cdot V_2(t) + \mu \cdot V_5(t) - (\lambda_2 + \mu + \lambda_s) \cdot V_6(t) + \lambda_2 \cdot V_8(t) \\
\frac{dV_7(t)}{dt} &= C_{op} + C_r \cdot \mu + \lambda_s \cdot V_3(t) + \mu \cdot V_5(t) - (\lambda_1 + \mu + \lambda_s) \cdot V_7(t) + \lambda_1 \cdot V_8(t) \\
\frac{dV_8(t)}{dt} &= C_p + 1/2C_r \cdot \mu + \lambda_s \cdot V_4(t) + 1/2\mu \cdot V_5(t) - (1/2\mu + \lambda_s) \cdot V_8(t)
\end{aligned} \quad (4.22)$$

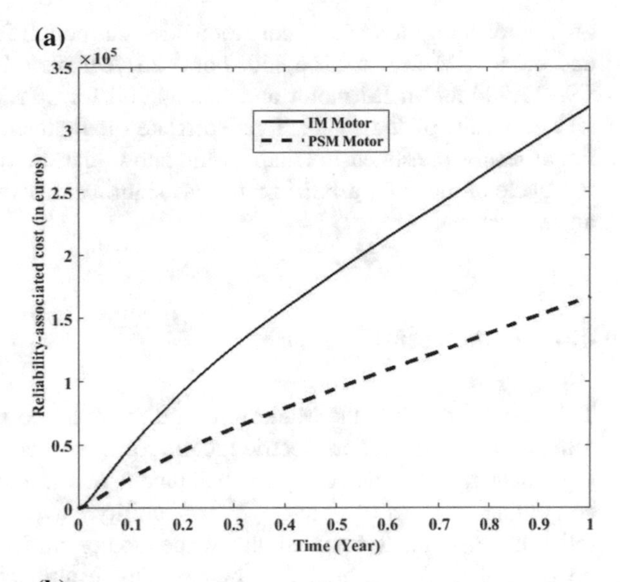

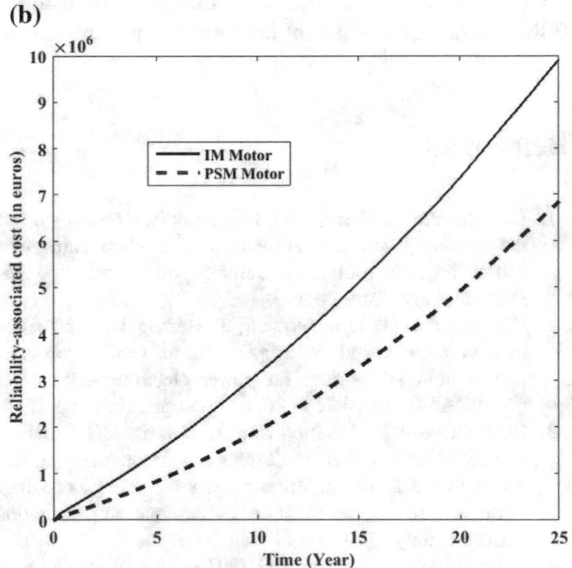

Fig. 4.8 Comparison of the MSS's reliability-associated cost with an IM motor (**a**) and a PSM motor (**b**)

Calculations were performed using the following costs:

- The system operating cost $C_{op} = 50{,}000$ euros per year
- The repair cost $C_r = 35{,}000$ euros per repair
- The penalty cost $C_p = 55{,}000$ euros per day

The results of the calculation can be seen in Fig. 4.8a, b.

The reliability-associated cost for a one-year period for an IM motor is €318,070 and for a PSM motor €166,860. For a 25 year-period of operations, the RAC is €9,911,900 for an IM motor and €6,834,000 for a PSM motor.

The results of the calculations correlate qualitatively with the statistical operational results presented in Chap. 2 and show that, if we take into account the full life cycle of the ship, a PSM motor has significant advantages in comparison with an IM motor.

4.4 Conclusion

This chapter presents the Markov reward approach to the comparative analysis of different types of traction electric motors for hybrid-electric propulsion systems. In this chapter, we discussed the development of the Markov reward approach for computing the MSS's average availability, average converted power and reliability-associated cost, all the while taking into consideration the vehicular operational conditions of electric motors. The results of the calculations show that a PSM motor has significant advantages in comparison with an IM motor.

References

1. Bolvashenkov I, Herzog HG, Rubinraut A, Romanovskiy V (2014) Possible ways to improve the efficiency and competitiveness of modern ships with electric propulsion systems. In: Proceedings of 10th IEEE vehicle power and propulsion conference (VPPC), Coimbra, Portugal, Nov 2014, pp 1–9
2. Bolvashenkov I, Kammermann J, Herzog HG (2016) Methodology for selecting electric traction motors and its application to vehicle propulsion systems. In: Proceedings of international symposium on power electronics, electrical drives, automation and motion (SPEEDAM), 22–24 June 2016, Anacapri, Italy, pp 1214–1219
3. Bolvashenkov I, Kammermann J, Herzog HG (2016) Research on reliability and fault tolerance of traction multi-phase permanent magnet synchronous motors based on Markov-models for multi-state systems. In: Proceedings of international symposium on power electronics, electrical drives, automation and motion, (SPEEDAM), 22–24 June 2016, Anacapri, Italy, pp 1166–1171
4. Bolvashenkov I, Herzog HG (2016) Use of stochastic models for operational efficiency analysis of multi power source traction drives. In: Frenkel I, Lisnianski A (eds) Proceedings of the second international symposium on stochastic models in reliability engineering, life science and operations management (SMRLO16), Beer Sheva, Israel, 2016, pp 124–130
5. Chu WQ, Zhu Z, Zhang J et al (2014) Comparison of electrically excited and interior permanent magnet machines for hybrid electric vehicle application. In: Proceedings of 17th international conference on electrical machines and systems (ICEMS), 2014, pp 401–407
6. Frenkel I, Bolvashenkov I, Herzog HG, Khvatskin L (2017) Operational sustainability assessment of multi power source traction drive. In: Ram M, Davim JP (eds) Mathematics applied to engineering. Elsevier, London, pp 191–203
7. Frenkel I, Khvatskin L, Lisnianski A (2010) Markov reward model for performance deficiency calculation of refrigeration system. In: Bris R, Soares CG, Martorell S

(eds) Reliability, risk and safety: theory and applications. CRC Press, Taylor & Francis Group, London, pp 1591–1596
8. Hashemnia N, Asaei B (2008) Comparative study of using different electric motors in electric vehicles. In: Proceedings of international conference on electrical machines (ICEM), 2008, pp 1–5
9. Howard R (1960) Dynamic programming and markov processes. MIT Press, Cambridge, Massachusetts
10. Kammermann J, Bolvashenkov I, Herzog HG (2015) Approach for comparative analysis of electric traction machines. In: Proceedings of 3rd international conference on electrical systems for aircraft, railway, ship propulsion, and road vehicles (ESARS) Aachen, Germany, Mar 2015, pp 1–5
11. Lisnianski A (2007) The Markov reward model for a multi-state system reliability assessment with variable demand. Qual Technol Qual Manag 4(2):265–278
12. Lisnianski A, Frenkel I (2009) Non-homogeneous Markov reward model for aging multi-state system under minimal repair. Int J Performability Eng 5(4):303–312
13. Lisnianski A, Frenkel I, Ding Y (2010) Multi-state system reliability analysis and optimization for engineers and industrial managers. Springer, London
14. Lisnianski A, Frenkel I, Khvatskin L, Ding Y (2008) Markov reward model for multi-state system reliability assessment. In: Vonta F, Nikulin M, Limnios N, Huber-Carol C (eds) Statistical models and methods for biomedical and technical systems. Birkhauser, Boston, pp 153–168
15. Menis R, da Rin A, Vicenzutti A, Sulligoi G (2012) Dependable design of all electric ships integrated power system: guidelines for system decomposition and analysis. In: Proceedings of electrical systems for aircraft, railway and ship propulsion (ESARS), Bologna, Italy, Oct 2012, pp 1–6
16. Mitra A, Emadi A (2012) On the suitability of large switched reluctance machines for propulsion applications. In: Proceedings of IEEE transportation electrification conference and expo (ITEC), 2012, pp 1–5
17. Postiglione CS, Collier DAF, Dupczak BS et al (2012) Propulsion system for an all electric passenger boat employing permanent magnet synchronous motors and modern power electronics. In: Proceedings of electrical systems for aircraft, railway and ship propulsion (ESARS), Bologna, Italy, Oct 2012, pp 1–6
18. Veneri O, Migliardini F, Capasso C, Corbo P (2012) Overview of electric propulsion and generation architectures for naval applications. In: Proceedings of electrical systems for aircraft, railway and ship propulsion (ESARS), Bologna, Italy, Oct 2012, pp 1–6
19. Wang W, Fahimi B (2012) Comparative study of electric drives for EV/HEV propulsion systems. In: Proceedings of IEEE transportation electrification conference and expo (ITEC), 2012, pp 1–6

Appendix
MATLAB® Codes for Chaps. 2 and 4

A.1 Using MATLAB® ODE Solvers

For the solution of systems of differential equations, MATLAB® provides functions, called *solvers*, which implement Runge-Kutta methods with variable step size. These functions are ode23, ode45 and ode113. In our calculations we have used the ode45 function, which employs a combination of fourth- and fifth-order methods and which is fast and accurate.

This function was used to solve the vector differential equation $\dot{\mathbf{p}} = \mathbf{f}(t, \mathbf{p})$ specified in the function file pdot, whose inputs must be t and p and whose output must be a *column* vector, representing $d\mathbf{p}/dt$; that is, $\mathbf{f}(t, \mathbf{p})$. The number of rows in this column vector must equal the order of the equations. The vector tspan contains the starting and ending values of the independent variable t. The vector p0 contains $\mathbf{p}(t_0)$. The basic syntax is:

$$[\mathrm{t}, \mathrm{p}] = \mathrm{ode45}(@\mathrm{funcpdot}, \mathrm{tspan}, \mathrm{p0})$$

I. Bolvashenkov et al., *Safety-Critical Electrical Drives*, SpringerBriefs in Electrical and Computer Engineering, https://doi.org/10.1007/978-3-319-89969-5

A.2 MATLAB® Cod for Reliability Indices Calculation of Multi-Power Source Traction Drives

A.2.1 MATLAB® Cod for System's Elements Definition

```
%-------------------------------
%            Amguema-type ship
%-------------------------------


function f=funcDE(t,p)
global L_DE Mu_DE;
f=zeros(2,1);
f(1) = -L_DE*p(1)+Mu_DE*(p(2));
f(2) = L_DE*p(1)-Mu_DE*p(2);

function f=funcG(t,p)
global L_G Mu_G;
f=zeros(2,1);
f(1) = -L_G*p(1)+Mu_G*(p(2));
f(2) = L_G*p(1)-Mu_G*p(2);

function f=funcMSb(t,p)
global L_MSb Mu_MSb;
f=zeros(2,1);
f(1) = -L_MSb*p(1)+Mu_MSb*(p(2));
f(2) = L_MSb*p(1)-Mu_MSb*p(2);

function f=funcEEC(t,p)
global L_EEC Mu_EEC;
f=zeros(2,1);
f(1) = -L_EEC*p(1)+Mu_EEC*(p(2));
f(2) = L_EEC*p(1)-Mu_EEC*p(2);

function f=funcEM1(t,p)
global L_EM1 Mu_EM1;
f=zeros(2,1);
f(1) = -L_EM1*p(1)+Mu_EM1*(p(2));
f(2) = L_EM1*p(1)-Mu_EM1*p(2);

function f=funcEM2(t,p)
global L_EM2 Mu_EM2;
f=zeros(2,1);
f(1) = -L_EM2*p(1)+Mu_EM2*(p(2));
f(2) = L_EM2*p(1)-Mu_EM2*p(2);

function f=funcFPP(t,p)
global L_FPP Mu_FPP;
f=zeros(2,1);
f(1) = -L_FPP*p(1)+Mu_FPP*(p(2));
f(2) = L_FPP*p(1)-Mu_FPP*p(2);

%-------------------------------
%            Norilsk-type ship
%-------------------------------


function f=funcMSDE1(t,p)
global L_MSDE1 Mu_MSDE1;
```

```
f=zeros(2,1);
f(1) = -L_MSDE1*p(1)+Mu_MSDE1*(p(2));
f(2) = L_MSDE1*p(1)-Mu_MSDE1*p(2);

function f=funcMSDE2(t,p)
global L_MSDE2 Mu_MSDE2;
f=zeros(2,1);
f(1) = -L_MSDE2*p(1)+Mu_MSDE2*(p(2));
f(2) = L_MSDE2*p(1)-Mu_MSDE2*p(2);

function f=funcFC1(t,p)
global L_FC1 Mu_FC1;
f=zeros(2,1);
f(1) = -L_FC1*p(1)+Mu_FC1*(p(2));
f(2) = L_FC1*p(1)-Mu_FC1*p(2);

function f=funcFC2(t,p)
global L_FC2 Mu_FC2;
f=zeros(2,1);
f(1) = -L_FC2*p(1)+Mu_FC2*(p(2));
f(2) = L_FC2*p(1)-Mu_FC2*p(2);

function f=funcCL1(t,p)
global L_CL1 Mu_CL1;
f=zeros(2,1);
f(1) = -L_CL1*p(1)+Mu_CL1*(p(2));
f(2) = L_CL1*p(1)-Mu_CL1*p(2);

function f=funcCL2(t,p)
global L_CL2 Mu_CL2;
f=zeros(2,1);
f(1) = -L_CL2*p(1)+Mu_CL2*(p(2));
f(2) = L_CL2*p(1)-Mu_CL2*p(2);

function f=funcGB(t,p)
global L_GB Mu_GB;
f=zeros(2,1);
f(1) = -L_GB*p(1)+Mu_GB*(p(2));
f(2) = L_GB*p(1)-Mu_GB*p(2);

function f=funcVPP(t,p)
global L_VPP Mu_VPP;
f=zeros(2,1);
f(1) = -L_VPP*p(1)+Mu_VPP*(p(2));
f(2) = L_VPP*p(1)-Mu_VPP*p(2);
```

A.2.2 MATLAB® Cod for Lz-Transform for Multi-state System

```
%-------------------------------
%            Amguema-type ship
%-------------------------------

%Variables definition

global L_DE Mu_DE;
global L_G Mu_G;
global L_MSb Mu_MSb;
global L_EEC Mu_EEC;
global L_EM1 Mu_EM1;
global L_EM2 Mu_EM2;
global L_FPP Mu_FPP;

%Variables initialization

L_DE = 2.2;
L_G = 0.15;
L_MSb = 0.05;
L_EEC = 0.2;
L_EM1 = 0.26;
L_EM2 = 0.26;
L_FPP = 0.01;

Mu_DE = 73.0;
Mu_G = 175.2;
Mu_MSb = 584.0;
Mu_EEC = 673.0;
Mu_EM1 = 117.0;
Mu_EM2 = 117.0;
Mu_FPP = 125.0;

ttt=[0:0.01:1];

p0_DE = [1 0];
[t_DE, p_DE] = ode45(@funcDE, ttt, p0_DE);
p0_G = [1 0];
[t_G, p_G] = ode45(@funcG, ttt, p0_G);
p0_MSb = [1 0];
[t_MSb, p_MSb] = ode45(@funcMSb, ttt, p0_MSb);
p0_EEC = [1 0];
[t_EEC, p_EEC] = ode45(@funcEEC, ttt, p0_EEC);
p0_EM1 = [1 0];
[t_EM1, p_EM1] = ode45(@funcEM1, ttt, p0_EM1);
p0_EM2 = [1 0];
[t_EM2, p_EM2] = ode45(@funcEM2, ttt, p0_EM2);
p0_FPP = [1 0];
[t_FPP, p_FPP] = ode45(@funcFPP, ttt, p0_FPP);

DE_1 = p_DE(:,1);
DE_2 = p_DE(:,2);
G_1 = p_G(:,1);
G_2 = p_G(:,2);
MSb_1 = p_MSb(:,1);
MSb_2 = p_MSb(:,2);
```

```
EEC_1 = p_EEC(:,1);
EEC_2 = p_EEC(:,2);
EM1_1 = p_EM1(:,1);
EM1_2 = p_EM1(:,2);
EM2_1 = p_EM2(:,1);
EM2_2 = p_EM2(:,2);
FPP_1 = p_FPP(:,1);
FPP_2 = p_FPP(:,2);

DG_1 = DE_1.*G_1;
DG_2 = DE_1.*G_2+DE_2;

DGS_1=DG_1.*DG_1.*DG_1.*DG_1;
DGS_2=4*DG_1.*DG_1.*DG_1.*DG_2;
DGS_3=6*DG_1.*DG_1.*DG_2.*DG_2;
DGS_4=4*DG_1.*DG_2.*DG_2.*DG_2;
DGS_5=DG_2.*DG_2.*DG_2.*DG_2;
DGS_SUM=DGS_1+DGS_2+DGS_3+DGS_4+DGS_5;

EM_1 = EM1_1.*EM2_1;
EM_2 = EM1_1.*EM2_2+EM1_2.*EM2_1;
EM_3 = EM1_2.*EM2_2;

SS1_1 = MSb_1.*DGS_1;
SS1_2 = MSb_1.*DGS_2;
SS1_3 = MSb_1.*DGS_3;
SS1_4 = MSb_1.*DGS_4;
SS1_5 = MSb_1.*DGS_5+MSb_2.*DGS_1+MSb_2.*DGS_2+MSb_2.*DGS_3
+MSb_2.*DGS_4+MSb_2.*DGS_5;

SS2_1 = EEC_1.*SS1_1;
SS2_2 = EEC_1.*SS1_2;
SS2_3 = EEC_1.*SS1_3;
SS2_4 = EEC_1.*SS1_4;
SS2_5 = EEC_1.*SS1_5+EEC_2.*SS1_1+EEC_2.*SS1_2+EEC_2.*SS1_3
+EEC_2.*SS1_4+EEC_2.*SS1_5;

SS3_1 = EM_1.*SS2_1;
SS3_2 = EM_1.*SS2_2;
SS3_3 = EM_1.*SS2_3+EM_2.*SS2_1+EM_2.*SS2_2+EM_2.*SS2_3;
SS3_4 = EM_1.*SS2_4+EM_2.*SS2_4;
SS3_5 = EM_1.*SS2_5+EM_2.*SS2_5+EM_3.*SS2_1+EM_3.*SS2_2
+EM_3.*SS2_3+EM_3.*SS2_4+EM_3.*SS2_5;

DED_1 = FPP_1.*SS3_1;
DED_2 = FPP_1.*SS3_2;
DED_3 = FPP_1.*SS3_3;
DED_4 = FPP_1.*SS3_4;
DED_5 = FPP_1.*SS3_5+FPP_2.*SS3_1+FPP_2.*SS3_2+FPP_2.*SS3_3
+FPP_2.*SS3_4+FPP_2.*SS3_5;
```

```
%--------------------------------
%           Norilsk-type ship
%--------------------------------

%Variables definition

global L_MSDE1 Mu_MSDE1;
global L_MSDE2 Mu_MSDE2;
global L_FC1 Mu_FC1;
global L_FC2 Mu_FC2;
global L_CL1 Mu_CL1;
global L_CL2 Mu_CL2;
global L_GB Mu_GB;
global L_VPP Mu_VPP;

%Variables initialization

L_MSDE1 = 1.5;
L_MSDE2 = 1.5;
L_FC1 = 0.15;
L_FC2 = 0.15;
L_CL1 = 0.11;
L_CL2 = 0.11;
L_GB = 0.11;
L_VPP = 0.1;
Mu_MSDE1 = 38.0;
Mu_MSDE2 = 38.0;
Mu_FC1 = 398.0;
Mu_FC2 = 398.0;
Mu_CL1 = 467.0;
Mu_CL2 = 467.0;
Mu_GB = 195.0;
Mu_VPP = 92.0;

ttt=[0:0.01:1];

p0_MSDE1 = [1 0];
[t_MSDE1, p_MSDE1] = ode45(@funcMSDE1, ttt, p0_MSDE1);
p0_MSDE2 = [1 0];
[t_MSDE2, p_MSDE2] = ode45(@funcMSDE2, ttt, p0_MSDE2);
p0_FC1 = [1 0];
[t_FC1, p_FC1] = ode45(@funcFC1, ttt, p0_FC1);
p0_FC2 = [1 0];
[t_FC2, p_FC2] = ode45(@funcFC2, ttt, p0_FC2);
p0_CL1 = [1 0];
[t_CL1, p_CL1] = ode45(@funcCL1, ttt, p0_CL1);
p0_CL2 = [1 0];
[t_CL2, p_CL2] = ode45(@funcCL2, ttt, p0_CL2);
p0_GB = [1 0];
[t_GB, p_GB] = ode45(@funcGB, ttt, p0_GB);
p0_VPP = [1 0];
[t_VPP, p_VPP] = ode45(@funcVPP, ttt, p0_VPP);


MSDE1_1 = p_MSDE1(:,1);
MSDE1_2 = p_MSDE1(:,2);
MSDE2_1 = p_MSDE2(:,1);
MSDE2_2 = p_MSDE2(:,2);
FC1_1 = p_FC1(:,1);
```

```
FC1_2 = p_FC1(:,2);
FC2_1 = p_FC2(:,1);
FC2_2 = p_FC2(:,2);
CL1_1 = p_CL1(:,1);
CL1_2 = p_CL1(:,2);
CL2_1 = p_CL2(:,1);
CL2_2 = p_CL2(:,2);
GB_1 = p_GB(:,1);
GB_2 = p_GB(:,2);
VPP_1 = p_VPP(:,1);
VPP_2 = p_VPP(:,2);

SS11_1 = MSDE1_1.*FC1_1;
SS11_2 = MSDE1_1.*FC1_2+MSDE1_2.*FC1_1+MSDE1_2.*FC1_2;

SS12_1 = CL1_1.*SS11_1;
SS12_2 = CL1_1.*SS11_2+CL1_2.*SS11_1+CL1_2.*SS11_2;

SS21_1 = MSDE2_1.*FC2_1;
SS21_2 = MSDE2_1.*FC2_2+MSDE2_2.*FC2_1+MSDE2_2.*FC2_2;

SS22_1 = CL2_1.*SS21_1;
SS22_2 = CL2_1.*SS21_2+CL2_2.*SS21_1+CL2_2.*SS21_2;

SS3_1 = SS12_1.*SS22_1;
SS3_2 = SS12_1.*SS22_2+SS12_2.*SS22_1;
SS3_3 = SS12_2.*SS22_2;

SS4_1 = GB_1.*SS3_1;
SS4_2 = GB_1.*SS3_2;
SS4_3 = GB_1.*SS3_3+GB_2.*SS3_1+GB_2.*SS3_2+GB_2.*SS3_3;

DGD_1 = VPP_1.*SS4_1;
DGD_2 = VPP_1.*SS4_2;
DGD_3 = VPP_1.*SS4_3+VPP_2.*SS4_1+VPP_2.*SS4_2+VPP_2.*SS4_3;
```

A.2.3 MATLAB® Cod for Availability, Performance and Performance Deficiency Calculation

```
%Availability calculation

A_DED_75=DED_1+DED_2;
A_DED_50=DED_1+DED_2+DED_3;

A_DGD_75=DGD_1;
A_DGD_50=DGD_1+DGD_2;

plot(ttt,A_DED_75,'k-',ttt,A_DED_50,'k--',ttt,A_DGD_75,
'r-',ttt,A_DGD_50,'r--','LineWidth',2);

%ylabel('Availability','FontSize',12,'FontWeight',
'bold','FontName','Times New Roman');

%Performance calculation

E_DED=5500*DED_1+4125*DED_2+2750*DED_3+1325*DED_4;
E_DGD=15440*DGD_1+7720*DGD_2;

%plot(ttt,E_DED,'k-',ttt,E_DGD,'k--','LineWidth',2);
%plot(ttt,E_DED/5500,'k-',ttt,E_DGD/15440,'k--','LineWidth',2);
%ylabel('Power Performance kW','FontSize',12,'FontWeight','bold',
'FontName','Times New Roman');
%ylabel('Relative Power Performance','FontSize',12,'FontWeight',
'bold','FontName','Times New Roman');

% Performance Deficiency calculation

D_DED_Winter=1375*DED_3+2750*DED_4+4125*DED_5;
D_DGD_Winter=3860*DGD_2+11580*DGD_3;
D_DED_Summer=1375*DED_4+2750*DED_5;
D_DGD_Summer=7720*DGD_3;

%plot(ttt,D_DED_Winter,'k-',ttt,D_DED_Summer,'k--','LineWidth',2);
%plot(ttt,D_DGD_Winter,'k-',ttt,D_DGD_Summer,'k--','LineWidth',2);
%ylabel('Power Performance Deficiency kW','FontSize',12,
'FontWeight','bold','FontName','Times New Roman');
xlabel('Time (Year)','FontSize',12,'FontWeight','bold','FontName',
'Times New Roman');
set(gca,'FontSize',12,'FontWeight','bold','FontName',
'Times New Roman');
```

A.3 MATLAB® Cod for the Markov Reward Approach for Selecting a Traction Electric Motor

A.3.1 MATLAB® Cod for the MSS's Average Availability Calculation

```
%Function - Average Availability Calculation for IM Motor for 2,750
kW demand

function f=funcMotors_IM_AvAv_2750(t,V)

global LambdaIM MuIM;

f=zeros(4,1);

Lambda1_IMt=LambdaIM+0.05*t;
Lambda2_IMt=LambdaIM+0.05*t;

f(1)=1-
(Lambda1_IMt+Lambda2_IMt)*V(1)+Lambda1_IMt*V(2)+Lambda2_IMt*V(3);
f(2)=1+MuIM*V(1)-(Lambda2_IMt+MuIM)*V(2)+Lambda2_IMt*V(4);
f(3)=1+MuIM*V(1)-(Lambda1_IMt+MuIM)*V(3)+Lambda1_IMt*V(4);
f(4)=0.5*MuIM*V(1)-0.5*MuIM*V(4);

%Function - Average availability calculation for IM Motor for 5,500
kW demand

function f=funcMotors_IM_AvAv_5500(t,V)

global LambdaIM MuIM;

f=zeros(4,1);

Lambda1_IMt=LambdaIM+0.05*t;
Lambda2_IMt=LambdaIM+0.05*t;

f(1)=1-
(Lambda1_IMt+Lambda2_IMt)*V(1)+Lambda1_IMt*V(2)+Lambda2_IMt*V(3);
f(2)=MuIM*V(1)-(Lambda2_IMt+MuIM)*V(2)+Lambda2_IMt*V(4);
f(3)=MuIM*V(1)-(Lambda1_IMt+MuIM)*V(3)+Lambda1_IMt*V(4);
f(4)=0.5*MuIM*V(1)-0.5*MuIM*V(4);
```

%Function - Average Availability Calculation for IM Motor for variable demand

```
function f=funcMotors_IM_AvAv_Seasonal(t,V)

global LambdaIM MuIM;
global LambdaW LambdaS;

f=zeros(8,1);

Lambda1_IMt=LambdaIM+0.05*t;
Lambda2_IMt=LambdaIM+0.05*t;

f(1)=1-(Lambda1_IMt+Lambda2_IMt+LambdaW)*V(1)+Lambda1_IMt*V(2)
     +Lambda2_IMt*V(3)+LambdaW*V(5);
f(2)=MuIM*V(1)-(Lambda2_IMt+MuIM+LambdaW)*V(2)+Lambda2_IMt*V(4)
     +LambdaW*V(6);
f(3)=MuIM*V(1)-(Lambda1_IMt+MuIM+LambdaW)*V(3)+Lambda1_IMt*V(4)
     +LambdaW*V(7);
f(4)=0.5*MuIM*V(1)-(0.5*MuIM+LambdaW)*V(4)+LambdaW*V(8);
f(5)=1+LambdaS*V(1)-(Lambda1_IMt+Lambda2_IMt+LambdaS)*V(5)
     +Lambda1_IMt*V(6) +Lambda2_IMt*V(7);
f(6)=1+LambdaS*V(2)+MuIM*V(5)-(Lambda2_IMt+MuIM+LambdaS)*V(6)
     +Lambda2_IMt*V(8);
f(7)=1+LambdaS*V(3)+MuIM*V(5)-(Lambda1_IMt+MuIM+LambdaS)*V(7)
     +Lambda1_IMt*V(8);
f(8)=LambdaS*V(4)+0.5*MuIM*V(5)-(0.5*MuIM+LambdaS)*V(8);
```

%Function - Average availability calculation for PSM Motor for 2,750 kW demand

```
function f=funcMotors_PSM_AvAv_2750(t,V)

global LambdaPSM MuPSM;

f=zeros(4,1);
Lambda1_PSMt=LambdaPSM+0.05*t;
Lambda2_PSMt=LambdaPSM+0.05*t;
f(1)=1-
(Lambda1_PSMt+Lambda2_PSMt)*V(1)+Lambda1_PSMt*V(2)+Lambda2_PSMt*V(3);
f(2)=1+MuPSM*V(1)-(Lambda2_PSMt+MuPSM)*V(2)+Lambda2_PSMt*V(4);
f(3)=1+MuPSM*V(1)-(Lambda1_PSMt+MuPSM)*V(3)+Lambda1_PSMt*V(4);
f(4)=0.5*MuPSM*V(1)-0.5*MuPSM*V(4);
```

%Function - Average availability calculation for PSM Motor for 5,500 kW demand

```
function f=funcMotors_PSM_AvAv_5500(t,V)

global LambdaPSM MuPSM;

f=zeros(4,1);
Lambda1_PSMt=LambdaPSM+0.05*t;
Lambda2_PSMt=LambdaPSM+0.05*t;
```

```
f(1)=1-
(Lambda1_PSMt+Lambda2_PSMt)*V(1)+Lambda1_PSMt*V(2)+Lambda2_PSMt*V(3);
f(2)=MuPSM*V(1)-(Lambda2_PSMt+MuPSM)*V(2)+Lambda2_PSMt*V(4);
f(3)=MuPSM*V(1)-(Lambda1_PSMt+MuPSM)*V(3)+Lambda1_PSMt*V(4);
f(4)=0.5*MuPSM*V(1)-0.5*MuPSM*V(4);
```

%Function - Average Availability Calculation for PSM Motor for variable demand

```
function f=funcMotors_PSM_AvAv_Seasonal(t,V)

global LambdaPSM MuPSM;
global LambdaW LambdaS;

f=zeros(8,1);

Lambda1_PSMt=LambdaPSM+0.05*t;
Lambda2_PSMt=LambdaPSM+0.05*t;

f(1)=1-(Lambda1_PSMt+Lambda2_PSMt+LambdaW)*V(1)+Lambda1_PSMt*V(2)
      +Lambda2_PSMt*V(3)+LambdaW*V(5);
f(2)=MuPSM*V(1)-(Lambda2_PSMt+MuPSM+LambdaW)*V(2)+Lambda2_PSMt*V(4)
      +LambdaW*V(6);
f(3)=MuPSM*V(1)-(Lambda1_PSMt+MuPSM+LambdaW)*V(3)+Lambda1_PSMt*V(4)
      +LambdaW*V(7);
f(4)=0.5*MuPSM*V(1)-(0.5*MuPSM+LambdaW)*V(4)+LambdaW*V(8);
f(5)=1+LambdaS*V(1)-(Lambda1_PSMt+Lambda2_PSMt+LambdaS)*V(5)
      +Lambda1_PSMt*V(6)+Lambda2_PSMt*V(7);
f(6)=1+LambdaS*V(2)+MuPSM*V(5)-(Lambda2_PSMt+MuPSM+LambdaS)*V(6)
      +Lambda2_PSMt*V(8);
f(7)=1+LambdaS*V(3)+MuPSM*V(5)-(Lambda1_PSMt+MuPSM+LambdaS)*V(7)
      +Lambda1_PSMt*V(8);
f(8)=LambdaS*V(4)+0.5*MuPSM*V(5)-(0.5*MuPSM+LambdaS)*V(8);
```

%Solver - Markov Reward Model for Average Availability Calculation

```
global LambdaIM MuIM;
global LambdaPSM MuPSM;
global LambdaW LambdaS;

LambdaIM=0.92;
LambdaPSM=0.39;
MuIM=60.8;
MuPSM=60.8;
LambdaW=5;
LambdaS=1.25;

ttt=[0:0.1:25];

V0_Motors_IM_AvAv_5500=[0 0 0 0];
[t_Motors_IM_AvAv_5500,V_Motors_IM_AvAv_5500]=ode45(@funcMotors_IM_Av
Av_5500, ttt, V0_Motors_IM_AvAv_5500);

V0_Motors_IM_AvAv_2750=[0 0 0 0];
[t_Motors_IM_AvAv_2750,V_Motors_IM_AvAv_2750]=ode45(@funcMotors_IM_Av
Av_2750, ttt, V0_Motors_IM_AvAv_2750);

V0_Motors_IM_AvAv_Seasonal=[0 0 0 0 0 0 0 0];
[t_Motors_IM_AvAv_Seasonal,V_Motors_IM_AvAv_Seasonal]=ode45(@funcMoto
rs_IM_AvAv_Seasonal, ttt, V0_Motors_IM_AvAv_Seasonal);
```

```
V0_Motors_PSM_AvAv_5500=[0 0 0 0];
[t_Motors_PSM_AvAv_5500,V_Motors_PSM_AvAv_5500]=ode45(@funcMotors_PSM
_AvAv_5500, ttt, V0_Motors_PSM_AvAv_5500);

V0_Motors_PSM_AvAv_2750=[0 0 0 0];
[t_Motors_PSM_AvAv_2750,V_Motors_PSM_AvAv_2750]=ode45(@funcMotors_PSM
_AvAv_2750, ttt, V0_Motors_PSM_AvAv_2750);

V0_Motors_PSM_AvAv_Seasonal=[0 0 0 0 0 0 0 0];
[t_Motors_PSM_AvAv_Seasonal,V_Motors_PSM_AvAv_Seasonal]=ode45(@funcMo
tors_PSM_AvAv_Seasonal, ttt, V0_Motors_PSM_AvAv_Seasonal);


A_Motors_IM_AvAv_5500=V_Motors_IM_AvAv_5500(:,1)./t_Motors_IM_AvAv_55
00;
A_Motors_IM_AvAv_2750=V_Motors_IM_AvAv_2750(:,1)./t_Motors_IM_AvAv_27
50;
A_Motors_IM_AvAv_Seasonal=V_Motors_IM_AvAv_Seasonal(:,1)./t_Motors_IM
_AvAv_Seasonal;

A_Motors_PSM_AvAv_5500=V_Motors_PSM_AvAv_5500(:,1)./t_Motors_PSM_AvAv
_5500;
A_Motors_PSM_AvAv_2750=V_Motors_PSM_AvAv_2750(:,1)./t_Motors_PSM_AvAv
_2750;
A_Motors_PSM_AvAv_Seasonal=V_Motors_PSM_AvAv_Seasonal(:,1)./t_Motors_
PSM_AvAv_Seasonal;

plot(ttt,A_Motors_IM_AvAv_5500, 'k--',ttt,A_Motors_IM_AvAv_2750, 'k-
.',
     ttt,A_Motors_IM_AvAv_Seasonal, 'k-','LineWidth',2);
plot(ttt,A_Motors_PSM_AvAv_5500(:,1), 'k—',
     ttt,A_Motors_PSM_AvAv_2750(:,1), 'k-.',
     ttt,A_Motors_PSM_AvAv_Seasonal(:,1), 'k-','LineWidth',2);

ylabel('Availability','FontSize',12,'FontWeight',
'bold','FontName','Times New Roman');
xlabel('Time (Year)','FontSize',12,'FontWeight',
'bold','FontName','Times New Roman');
set(gca,'FontSize',12,'FontWeight','bold','FontName','Times New
Roman');
```

A.3.2 MATLAB® Cod for the MSS's Average Converted Power Calculation

```
%Function - Average converted power calculation for IM Motor

function f=funcMotors_IM_AvPerf(t,V)

global LambdaIM MuIM;

f=zeros(4,1);

Lambda1_IMt=LambdaIM+0.05*t;
Lambda2_IMt=LambdaIM+0.05*t;

f(1)=5500-
(Lambda1_IMt+Lambda2_IMt)*V(1)+Lambda1_IMt*V(2)+Lambda2_IMt*V(3);
f(2)=2750+MuIM*V(1)-(Lambda2_IMt+MuIM)*V(2)+Lambda2_IMt*V(4);
f(3)=2750+MuIM*V(1)-(Lambda1_IMt+MuIM)*V(3)+Lambda1_IMt*V(4);
f(4)=0.5*MuIM*V(1)-0.5*MuIM*V(4);

%Function - Average converted power calculation for PSM Motor

function f=funcMotors_PSM_AvPerf(t,V)

global LambdaPSM MuPSM;

f=zeros(4,1);

Lambda1_PSMt=LambdaPSM+0.05*t;
Lambda2_PSMt=LambdaPSM+0.05*t;

f(1)=5500-
(Lambda1_PSMt+Lambda2_PSMt)*V(1)+Lambda1_PSMt*V(2)+Lambda2_PSMt*V(3);
f(2)=2750+MuPSM*V(1)-(Lambda2_PSMt+MuPSM)*V(2)+Lambda2_PSMt*V(4);
f(3)=2750+MuPSM*V(1)-(Lambda1_PSMt+MuPSM)*V(3)+Lambda1_PSMt*V(4);
f(4)=0.5*MuPSM*V(1)-0.5*MuPSM*V(4);

%Solver - Markov Reward Model for Average Converted Power calculation

global LambdaIM MuIM;
global LambdaPSM MuPSM;

LambdaIM=0.92;
LambdaPSM=0.39;

MuIM=60.8;
MuPSM=60.8;

ttt=[0:0.1:25];

V0_Motors_IM_AvPerf=[0 0 0 0];
[t_Motors_IM_AvPerf,V_Motors_IM_AvPerf]=ode45(@funcMotors_IM_AvPerf,
    ttt, V0_Motors_IM_AvPerf);

V0_Motors_PSM_AvPerf=[0 0 0 0];
[t_Motors_PSM_AvPerf,V_Motors_PSM_AvPerf]=ode45(@funcMotors_PSM_AvPer
f,ttt, V0_Motors_PSM_AvPerf);

E_Motors_IM_AvPerf=V_Motors_IM_AvPerf(:,1)./t_Motors_IM_AvPerf;
E_Motors_PSM_AvPerf=V_Motors_PSM_AvPerf(:,1)./t_Motors_PSM_AvPerf;

plot(ttt,E_Motors_IM_AvPerf(:,1), 'k-',
     ttt,E_Motors_PSM_AvPerf(:,1), 'k--','LineWidth',2);

ylabel(' Average Converted Power calculation, KW','FontSize',12,
       'FontWeight','bold','FontName','Times New Roman');
xlabel('Time (Year)','FontSize',12,'FontWeight','bold','FontName',
       'Times New Roman');
set(gca,'FontSize',12,'FontWeight','bold','FontName','Times New
Roman');
```

A.3.3 MATLAB® Cod for He MSS's Reliability-Associated Cost Calculation

```
%Function - Reliability-associated cost calculation for IM Motor

function f=funcMotors_IM_RAC(t,V)

global LambdaIM MuIM;
global LambdaW LambdaS;
global Cop Cr Cp;

f=zeros(8,1);

Lambda1_IMt=LambdaIM+0.05*t;
Lambda2_IMt=LambdaIM+0.05*t;


f(1)=Cop-(Lambda1_IMt+Lambda2_IMt+LambdaW)*V(1)+Lambda1_IMt*V(2)
        +Lambda2_IMt*V(3)+LambdaW*V(5);
f(2)=Cp+Cr*MuIM+MuIM*V(1)-(Lambda2_IMt+MuIM+LambdaW)*V(2)
       +Lambda2_IMt*V(4)+LambdaW*V(6);
f(3)=Cp+Cr*MuIM+MuIM*V(1)-(Lambda1_IMt+MuIM+LambdaW)*V(3)
       +Lambda1_IMt*V(4)+LambdaW*V(7);
f(4)=Cp+0.5*Cr*MuIM+0.5*MuIM*V(1)-
(0.5*MuIM+LambdaW)*V(4)+LambdaW*V(8);
f(5)=Cop+LambdaS*V(1)-(Lambda1_IMt+Lambda2_IMt+LambdaS)*V(5)
        +Lambda1_IMt*V(6)+Lambda2_IMt*V(7);
f(6)=Cop+Cr*MuIM+LambdaS*V(2)+MuIM*V(5)-
(Lambda2_IMt+MuIM+LambdaS)*V(6)
        +Lambda2_IMt*V(8);
f(7)=Cop+Cr*MuIM+LambdaS*V(3)+MuIM*V(5)-
(Lambda1_IMt+MuIM+LambdaS)*V(7)
        +Lambda1_IMt*V(8);
f(8)=Cp+0.5*Cr*MuIM+LambdaS*V(4)+0.5*MuIM*V(5)-
(0.5*MuIM+LambdaS)*V(8);

%Function - Reliability Associated Cost Calculation for PSM Motor

function f=funcMotors_PSM_RAC(t,V)

global LambdaPSM MuPSM;
global LambdaW LambdaS;
global Cop Cr Cp;

f=zeros(8,1);

Lambda1_PSMt=LambdaPSM+0.05*t;
Lambda2_PSMt=LambdaPSM+0.05*t;
```

```
f(1)=Cop-(Lambda1_PSMt+Lambda2_PSMt+LambdaW)*V(1)+Lambda1_PSMt*V(2)
       +Lambda2_PSMt*V(3)+LambdaW*V(5);
f(2)=Cp+Cr*MuPSM+MuPSM*V(1)-(Lambda2_PSMt+MuPSM+LambdaW)*V(2)
       +Lambda2_PSMt*V(4)+LambdaW*V(6);
f(3)=Cp+Cr*MuPSM+MuPSM*V(1)-(Lambda1_PSMt+MuPSM+LambdaW)*V(3)
       +Lambda1_PSMt*V(4)+LambdaW*V(7);
f(4)=Cp+0.5+Cr*MuPSM+0.5*MuPSM*V(1)-(0.5*MuPSM+LambdaW)*V(4)
       +LambdaW*V(8);
f(5)=Cop+LambdaS*V(1)-(Lambda1_PSMt+Lambda2_PSMt+LambdaS)*V(5)
       +Lambda1_PSMt*V(6)+Lambda2_PSMt*V(7);
f(6)=Cop+Cr*MuPSM+LambdaS*V(2)+MuPSM*V(5)-(Lambda2_PSMt+MuPSM
        +LambdaS)*V(6)+Lambda2_PSMt*V(8);
f(7)=Cop+Cr*MuPSM+LambdaS*V(3)+MuPSM*V(5)-(Lambda1_PSMt+MuPSM
        +LambdaS)*V(7)+Lambda1_PSMt*V(8);
f(8)=Cp+0.5*Cr*MuPSM+LambdaS*V(4)+0.5*MuPSM*V(5)
       -(0.5*MuPSM+LambdaS)*V(8);
```

%Solver - Markov Reward Model for Reliability-associated cost calculation

```
global LambdaIM MuIM;
global LambdaPSM MuPSM;
global LambdaW LambdaS;
global Cop Cr Cp;

LambdaIM=0.92;
LambdaPSM=0.39;

MuIM=60.8;
MuPSM=60.8;

LambdaW=5;
LambdaS=1.25;

Cop=50000;
Cr=35000;
Cp=20075000;
ttt=[0:0.01:1];


V0_Motors_IM_RAC=[0 0 0 0 0 0 0 0];
[t_Motors_IM_RAC,V_Motors_IM_RAC]=ode45(@funcMotors_IM_RAC,
    ttt, V0_Motors_IM_RAC);

V0_Motors_PSM_RAC=[0 0 0 0 0 0 0 0];
[t_Motors_PSM_RAC,V_Motors_PSM_RAC]=ode45(@funcMotors_PSM_RAC,
    ttt, V0_Motors_PSM_RAC);

plot(ttt,V_Motors_IM_RAC(:,1), 'k-',ttt,V_Motors_PSM_RAC(:,1),
    'k--','LineWidth',2);

ylabel('Reliability Associated Cost (EU)','FontSize',12,'FontWeight',
       'bold','FontName','Times New Roman');
xlabel('Time (Year)','FontSize',12,'FontWeight','bold','FontName',
       'Times New Roman');
set(gca,'FontSize',12,'FontWeight','bold','FontName','Times New
Roman');
```

Printed in the United States
By Bookmasters